English

The Language of Science and Innovation

How Latin Precision, Greek Philosophy, and Germanic Expressiveness Made English Powerful.

By

ENZO MILANO

English: The Language of Science and Innovation ©

Table of Contents

Introduction

The English language has risen to prominence as a global lingua franca, facilitating intellectual exchange across cultures and disciplines. However, English did not achieve this status arbitrarily. Its ascendancy is rooted in a rich linguistic history and influenced by the cultural forces that shaped its evolution.

This book delves into the factors that have established English as the preeminent language of science, philosophy, and technology. It explores how the language emerged at the intersection of diverse linguistic traditions, absorbing elements from Latin, Greek, and Germanic languages. Through this mixture of influences, English developed an unparalleled capacity for precision as well as emotive and narrative depth.

The following chapters examine how each contributing language left its mark. Latin infused English with an extensive vocabulary of specialized terms, favoring clarity and exactitude. Greek contributions brought an abundance of philosophical and scientific concepts, casting a wide net of abstract reasoning. Meanwhile, the English language drew structural foundations and an affinity for storytelling from its Germanic roots.

By amalgamating these varied traditions, English achieved a rare blend of attributes. It gained versatility across technical and creative applications while providing cognitive ease through recognizable linguistic patterns. Most significantly, English became uniquely positioned to serve as a conduit for ideas—synthesizing influences from ancient scholarship to contemporary innovations.

This book aims to shed light on the intricate cultural and historical processes that transformed English into an intellectual powerhouse. It traces how the language navigated shifts in Western thought by maintaining continuity with the past yet adapting to modern pursuits of knowledge. Through case studies and analyses, the following chapters will illustrate how English has consistently proven to be a catalyst for scientific and philosophical progress, earning its prestigious role as the foremost linguistic instrument of global exchange.

The modern English language emerged at a nexus of linguistic traditions that converged over centuries of cultural and political change in Europe. As the Roman Empire expanded across the continent, Latin became dominant in law, education and imperial administration. In Britain, the expansion brought Latin influences but the local Germanic tribes maintained their own language and culture. Over time, following the decline of Rome, Latin and Anglo-Saxon Old English fused into a new pluralistic tongue.

This initial blending of Latin and Germanic roots coincided with power shifts across Europe. As feudal kingdoms and empires rose and fell, English continued absorbing vocabulary from concurrent influences like Old Norse and Norman French. This infusion aligned with periods of cultural diffusion under new political configurations. By the Renaissance, English had evolved into a flexible medium for intellectuals rediscovering ancient Greek texts and experimenting with new ideas. The language readily adopted neologisms describing concepts in philosophy, science and the arts.

During the Age of Discovery, English spread globally through colonial expansion. However, its true global dominance emerged from associations with advances of the Western intellectual tradition. In realms like the Enlightenment, Industrial Revolution and Modern sciences, English cemented its prestige as the preferred language of technical literature and innovation. While other European tongues also comprised Latin, Greek and indigenous elements, English achieved a rare synthesis between precision and ambiguity. It retained Latin's clarity while embracing the ambiguity of Germanic constructs, facilitating interdisciplinary discourse.

As new academic disciplines emerged, English continued evolving to flexibly describe multidisciplinary frameworks. Fields like psychology, computer sciences, genetics and more influenced its lexicon. In this way, the language maintained continuity as a mirror of changing paradigms. Even today, it absorbs terminology from diverse international communities of scholars and experts. This allows English to remain the foremost language serving global exchange in an age of growing specialization and interconnectivity.

The following chapters will outline how English navigation of these intricate cultural and historical forces established it as the preeminent language of scientific progress. Specific case studies, theories and research will illustrate its unique facilitation of innovation across eras by deftly synthesizing influences of the past with terminology for present inquiries, thereby building the argument for why English achieved worldwide standing as the preeminent language of intellect.

English: The Language of Science and Innovation ©

Chapter 1

The Linguistic Crucible

Unraveling Language Diversity

The modern English language is often described as a linguistic melting pot, having emerged from the comingling of multiple influences over the centuries.

Through waves of migration, conquest and cultural diffusion, English absorbed vocabulary and constructs from Latin, Greek, French, Norse and other Germanic languages. "Unraveling Language Diversity" explores how each contribution enriched the expressive capacity of English. Vocabulary from Romance languages augmented precise terminology while Germanic roots enhanced narrative storytelling.

By amalgamating these varied components, English gained a versatility few other languages achieved. It could delicately describe abstract philosophical concepts as well as capturing the gritty realities of daily life. English's inherent tendency toward synthesis nurtured an inclusive linguistic framework. Within this framework, foreign words could be smoothly assimilated without displacement of native elements.

Importantly, the language's absorbing quality meant it could continuously represent new ideas regardless of origin. As discussed in "Exploring how the English language, as a melting pot of linguistic elements, fosters an environment for diverse ideas and concepts to converge," its inclusive nature made English uniquely suited to serve as a common medium. Diverse academic disciplines and international scholars could communicate through shared terminology.

English emerged as a remarkable crucible through centuries of contact with multitudes of languages and cultures. Its strategic positioning at the crossroads of linguistic commerce endowed English with rare adaptive powers supporting convergence instead of division. This crucible effect underlies why English became the preeminent language of global intellectual exchange.

English emerged as a linguistic crucible, a result of its exposure to various languages and cultures through centuries of contact. This exposure, particularly through waves of invasion and settlement, allowed English to absorb vocabulary and constructs from languages such as Latin, Greek, French, Old Norse, and other Anglo-Saxon influences.

The strategic positioning of English at the crossroads of linguistic commerce between continental Europe and the British Isles afforded it unique advantages compared to other Western European languages. English was not confined to a particular geographical region or limited to educational use, like Latin. Instead, it was exposed to constant interaction and exchange with outside linguistic currents, which led to the development of rare adaptive powers.

English's ability to synthesize foreign elements into its structure without threatening the integrity of its native roots is a result of its crucible effect. This process allowed new contributions to assimilate while retaining continuity with older influences, leading to a versatile inclusiveness that is scarce among other widely spoken languages.

As the linguistic crossroads of Britain, English became capable of representing a diverse range of ideas from global cultures. Its keywords and concepts could resonate broadly with international communities, regardless of their origins. This unique quality made English an ideal medium for universal intellectual exchange, which contributed to its rise to prominence on the global stage.

The accrued benefits of English's strategic localization and centuries of contact as a linguistic crucible underlie its emergence as a dominant language. This historical development has enabled English to serve as a lingua franca, facilitating communication and exchange among diverse cultures and communities.

Furthermore, the versatility of English has allowed it to adapt to various contexts, from scientific and academic discourse to popular culture and artistic expression. Its inclusive nature has also fostered the development of diverse dialects and hybrid languages, reflecting the diverse cultural influences that have shaped English over time.

English's emergence as a linguistic crucible is a result of its exposure to diverse languages and cultures, its adaptive powers, and its strategic positioning at the crossroads of linguistic commerce. These factors have contributed to English's unique ability to represent a broad range of ideas from global cultures, making it an ideal medium for universal intellectual exchange and a dominant language in modern times.

Historical Evolution of English

The previous section explored the cultural processes through which English emerged as a repository of diverse linguistic influences. This positioned the language as a unique crucible for synthesis and conceptual representation. Chapter 1 now shifts focus to the historical timeline along which English evolved into an unparalleled global lingua franca.

"Tracing the linguistic journey and the evolution that shaped English into a global medium for intellectual exchange" charts the progression of English over centuries. It describes the successive stages of Old English, Middle English and Early Modern English as new forms crystallized through ongoing cultural contact. Key historical junctures like the Norman Conquest of 1066 and impacts of the Renaissance are examined for their role in augmenting the language.

The chapter traces expansion of English usage from Britain across domains of law, education and imperial administration. It discusses the language's growing prestige in technical writings concurrent with strides of the European Enlightenment. English further spread through global colonization, cementing popularity in commerce and politics internationally.

Significantly, the section highlights English's unique ability to continuously represent new scientific revolutions throughout modern periods. From physics to genetics, the language assimilated specialist vocabulary to describe cutting-edge frontier knowledge with precision. This flexibility permitted English to assert premier status as the medium for worldwide intellectual exchange ever since.

Together, these segments map the rich historical lineage and fortuitous developmental path that established English as the principal global lingua franca of the present digital age.

The emergence of English as a global lingua franca is a remarkable story that spans centuries. From its humble beginnings as a local language in Britain, English has evolved into a versatile and dynamic tool for communication, accommodating diverse linguistic influences and cultural exchange. This chapter traces the historical timeline of English's growth and development, highlighting key milestones that contributed to its rise as a premier global language.

Old English (c. 450-1100)

The origins of English can be traced back to the 5th century, when Germanic tribes such as the Angles, Saxons, and Jutes invaded Britain. Their language, known as Old English, was heavily influenced by Latin, which was the language of the Roman Empire. This period saw the development of the English alphabet, as well as the establishment of a basic grammatical structure that would lay the foundation for future linguistic innovations.

Middle English (c. 1100-1500)

The Norman Conquest of 1066 marked a significant turning point in the evolution of English. The Norman French language, spoken by the ruling class, had a profound impact on English, leading to the borrowing of many French words and the development of a more complex vocabulary. This period also saw the emergence of Chaucerian English, which became the standard form of written English.

Early Modern English (c. 1500-1800)

The Renaissance brought about a renewed interest in classical learning, which had a significant impact on the English language. This period saw the introduction of many Latin and Greek words, as well as the standardization of English spelling and grammar. The rise of the British Empire also contributed to the spread of English across the globe, as the language became a tool for administration and commerce.

Modern English (c. 1800-present)

The Industrial Revolution marked a significant turning point in the evolution of English. The language became more accessible to the masses, and the emergence of new technologies and scientific discoveries led to the creation of new vocabulary. English also became a dominant language in international relations, business, and education, cementing its status as a global lingua franca.

Adaptability and Assimilation

One of the key factors that contributed to English's rise as a global language was its ability to adapt and assimilate new words and ideas. The language's capacity to absorb vocabulary from other languages, particularly Latin, Greek, and French, allowed it to expand its vocabulary and express complex ideas with precision. This flexibility also enabled English to accommodate new scientific and technological advancements, such as the terms "physics" and "genetics," which were adopted from Latin and Greek.

The evolution of English as a global lingua franca is a testament to the language's adaptability, flexibility, and capacity to represent new ideas and concepts. From its humble beginnings as a local language in Britain, English has emerged as a premier global language, capable of accommodating diverse linguistic influences and cultural exchange. The historical timeline traced in this chapter highlights the key milestones that contributed to English's rise, and underscores the language's unique ability to continuously represent new scientific revolutions throughout modern periods.

English: The Language of Science and Innovation ©

Chapter 2

The Cultural Fabric

The Influence of Western Thought

Analyzing the impact of Western ideologies on shaping intellectual pursuits and encouraging individualism.

Many of the ideologies that emerged from Western thought had a profound influence on the evolution of English and its role in intellectual pursuits. Beginning in the Renaissance, thinkers increasingly valorized principles of humanism, rationalism and individualism. This marked a shift away from rigid theological authority and towards endorsement of independent logical reasoning and empirical study of the natural world.

Concepts like these took root and flourished during the Enlightenment era. Philosophers like Locke, Voltaire and Hume advanced ideals of evidence-based thinking, questioning long-held traditions, and elevating the capability of the human mind to understand reality through observation and experimentation. Their works were widely disseminated and debated throughout English-speaking societies. Gradually, Western thought embraced democratic ethics and celebration of individual liberty as cornerstones of civil progress.

As these ideological currents gained momentum, English adapted linguistically to precisely articulate emerging frameworks. Neologisms entered describing new epistemological approaches and branches of scientific analysis. The vocabulary of English synthesized with philosophies encouraging broader participation in knowledge-generation as an avenue for personal development and societal betterment. Over time, English was increasingly associated with empowering independent reasoning rather than just relaying established concepts.

The language also synchronized with gradual liberalization of cultural norms. It embraced more ambiguous stylistic elements allowing flexible expression of subjective experience and progressive viewpoints. English thrived in environments valuing self-directed learning and individual expression of new hypotheses. Its flexibility complemented the open, questioning spirit championed by thinkers of the Enlightenment and continued encouraging dissent from entrenched positions.

In these ways, English became intertwined with the ideological transformation underway in the Western world, cementing its role as an indispensable instrument for intellectual change and development of new fields of human inquiry. The language synchronized with cultural trends prioritizing rationalism, empiricism and individualism in shaping new frontiers of scholarship.

Cultural Cross-Pollination

How the English language facilitates the fusion of diverse cultural concepts, enriching the intellectual landscape.

English has had a unique advantage as a linguistic conduit for cultural cross-pollination due to its status as a key lingua franca. As a language formed at the intersection of diverse linguistic influences, English developed a special capacity to smoothly assimilate foreign terminology from various cultural spheres. Over centuries, it absorbed vocabulary from global contact through avenues like exploration, trade, academia and migration.

The language retains a permeable quality allowing incorporation of new words without compromising its core integrity. Foreign lexical imports can be seamlessly anglicized to resonate with both native English speakers and global audiences. This cross-pollinating function has enriched English by continually enhancing its scope and precision. As new domains and technologies emerge, associated terminology rapidly streams into common usage from other cultural contexts through the language.

Certain periods especially epitomized English's role in global exchange of intellectual trends. During the Renaissance, rediscovery of ancient Greek philosophy and artistic works spurred an influx of Hellenic concepts. In the modern age, English serves as the primary medium for international scientific cooperation. Studies demonstrate how linguistic assimilation of neologisms from diverse cultures accelerates cross-disciplinary applications by integrating multiple perspectives into shared discourses.

English thrives as an avenues for blending ideas precisely due to its centuries-honed ability to represent an unparalleled diversity of cultural worldviews. Flexible integration rather than replacement of foreign influences mirrors wider trends of globalization and connectivity in the current digital landscape. In this way, English continues enriching the intellectual topography by catalyzing fusion rather than division between human perspectives and fields of learning.

Chapter 3

Philosophical Underpinnings

Free Thought and Inquiry

Exploring how English and Western ideals of free thinking stimulate philosophical contemplation and inquiry.

English evolved alongside intellectual movements in the West that championed principles of rationalism, empiricism and liberty of ideas. As these values took shape, the language adapted nuanced terminology to support emerging trends valuing independence of thought and experimentation. Words entered describing novel methodologies, theories and branches of questioning like scientific philosophy.

Over time, the vocabulary and stylistic elements of English began implicitly encouraging skepticism of established wisdoms and exploration of plural perspectives. As a medium increasingly used in democratic and individualistic societies, it came to embrace ambiguity allowing polite but rigorous challenging of viewpoints. Together with cultural shifts, this emboldened participation in philosophical debates across diverse public spheres.

English afforded flexibility for hypothesizing, while still enabling precise articulation of logical steps. As such, it became a favored instrument for pioneers profiling new lenses of inspection like during scientific revolutions. The language paralleled expanding endorsement of rational free thought as integral to societal and technological progress. Gradually, its constructs mirrored cultural shifts toward open-minded and self-guided learning at all strata of education as well.

Contemporary research analyzes correlation between linguistic traits and cognitive processes. Studies find English encourages inductive reasoning, consideration of alternatives and re-evaluation of default assumptions in argumentation due to its symbiosis with liberty of ideas. The language harmonized with transitioning intellectual climates that prioritized skepticism, experimentation and diversity of philosophical perspective as cornerstones of advancing human knowledge and standards of living.

The influence of English philosophers and thinkers such as Locke, Mill, and Russell has been instrumental in shaping English as a language of debate and inquiry. Their works have promoted rationalism and empiricism, which have encouraged the use of evidence and logical reasoning in intellectual discourse. This has helped to create a culture of critical thinking and open-mindedness, where individuals can engage in constructive debate and explore novel ideas without fear of censorship.

One historical example of the impact of English on scientific progress is the facilitation of scientific revolutions from Copernicus to Darwin. English allowed for the flexible discussion of novel theories, enabling scientists to present their ideas and engage in open debate, leading to the development of new scientific frameworks.

Linguistic elements such as hedging language have also played a crucial role in nurturing a questioning spirit in English. Hedging language encourages tentative and nuanced presentation of ideas, inviting feedback and discussion without fear of censure. This has created an environment where individuals can explore new ideas and perspectives without being judged or ridiculed.

Another significant aspect of English is its absorption of terms from diverse academic fields, enabling interdisciplinary connections and sparking new perspectives through hybridized frameworks. This cross-pollination of ideas has led to the development of new fields of study and has facilitated the exchange of knowledge across disciplines.

The decentralized and nonlinear structure of English compositions mirrors the open-ended and meandering nature of inquiry. It allows individuals to present their ideas and thoughts in a flexible and dynamic manner, reflecting the way we think and reason. This structure also encourages creativity and serendipitous discoveries, as individuals can explore new ideas and connections without being limited by a rigid framework.

Finally, English serves as an inclusive platform that accommodates marginalized or minority viewpoints. This diversity and cultural pluralism have fueled creativity and innovation, as individuals from different backgrounds and perspectives can come together to exchange ideas and collaborate. New media has further empowered individuals as citizen scholars globally, allowing them to engage in intellectual discourse and contribute to collective wisdom. English functions as their shared space to exchange perspectives and collaboratively advance knowledge.

The influence of English philosophers and thinkers, linguistic elements such as hedging language, the absorption of terms from diverse academic fields, the decentralized structure of English compositions, and the inclusive nature of the language have all contributed to English becoming a language of debate and inquiry. These factors have created an environment where individuals can engage in constructive debate, explore novel ideas, and collaborate to advance collective wisdom.

Individual Pursuit of Truth

Examining how the language and cultural backdrop encourage the quest for deeper truths and philosophical musings.

English evolved alongside Western ideals that valorized reason and evidence-based scrutiny over received wisdom. As these principles took hold, the language absorbed terminology reflecting new epistemological priorities and methodologies for advancing knowledge claims. Concepts like skepticism, empiricism and the scientific method entered common parlance.

Across eras like the Enlightenment, English adopted flexible structures mirroring increasing trust in an individual's capacity to deduce factual realities and philosophical insights through autonomous investigation. Hedges, hedges and other linguistic tools promote tentative conclusions, inviting open critique to strengthen or refine arguments. This nurtures an empirical spirit of continual re-evaluation.

English also remains permeable to specialized lexicons from diverse fields, facilitating interdisciplinary links. The fluid assimilation of foreign concepts sparks new frameworks that push inquiry in novel directions. Over time, the language came to intrinsically encourage intellectual risk-taking and creative hypothesizing instead of just relaying orthodoxies.

Further, English serves as an inclusive platform globally. It empowers diverse actors as citizen scholars freely pursuing philosophical discoveries through open exchange and collaborative experimentation online. Linguistic and cultural shifts have progressively supported intellectual independence and democratization of knowledge production.

English harmonized with trends valorizing autonomy, skepticism and evidence-based rationalism. The language intrinsically aids a questioning spirit and flexibility to integrate diverse lenses that advance both collective learning and satisfaction of individual curiosities. This nurtures continual progress on humanity's shared quest for philosophical and scientific truths.

English has evolved in tandem with Western ideals that prioritize reason, evidence, and critical thinking. This has resulted in the language absorbing terminology that reflects new epistemological priorities and methodologies for advancing knowledge claims. Concepts like skepticism, empiricism, and the scientific method have entered common parlance, and the language has adopted flexible structures that mirror the increasing trust in individuals' abilities to deduce factual realities and philosophical insights through autonomous investigation.

English has also become permeable to specialized lexicons from diverse fields, facilitating interdisciplinary links and sparking new frameworks that push inquiry in novel directions. The language has intrinsically encouraged intellectual risk-taking and creative hypothesizing, rather than just relaying orthodoxies.

Furthermore, English serves as an inclusive platform globally, empowering diverse actors as citizen scholars who can freely pursue philosophical discoveries through open exchange and collaborative experimentation online. Linguistic and cultural shifts have progressively supported intellectual independence and the democratization of knowledge production.

English has harmonized with trends that valorize autonomy, skepticism, and evidence-based rationalism. The language's flexibility and ability to integrate diverse lenses aid a questioning spirit and promote continual progress on humanity's shared quest for philosophical and scientific truths.

Chapter 4

Scientific Innovation

Language and Scientific Exploration

Investigating the role of English in propelling scientific inquiry and breakthroughs.

English has served as a highly effective linguistic tool for the dissemination and advancement of scientific knowledge. Its flexible structures and absorption of specialized terminology from various languages allowed precise articulation of innovative theories and findings.

Notable examples include how English facilitated revolutionary scientific works. Copernicus, Newton and Darwin all utilized the language's clarity and expressiveness to radically transform core models of astronomy, physics and biology. Their ability to present and defend novel hypotheses in English encouraged broader consideration and testing of ideas.

As the language of the Royal Society and other early scientific organizations, English became strongly associated with principles of empirical evidence and reproducibility. It synthesized structured methods of logical argumentation with an open, questioning spirit.

This symbiosis between language and method supported unprecedented collaborative progress across borders.

In the modern age, English remains the primary medium of global scientific publishing and conferences. Its worldwide reach and continuous acquisition of neologisms ensure concepts and data rapidly permeate networks of international experts. Linguistic equality of access encourages diverse teams to push frontiers in fields like medicine, genomics and renewable technology.

Studies also show English textual features like hedging promote tentative, nuanced theorizing that invites peer review strengthening hypotheses. The language's versatility, precision and inclusive qualities have seamlessly supported each major phase of scientific exploration and exchange of innovative discoveries over the centuries.

This has enabled scientific communities worldwide to share knowledge, collaborate, and accelerate progress in their respective fields. The language's far-reaching influence has facilitated the rapid dissemination of concepts, data, and innovative ideas across international networks of experts, fostering a culture of collaboration and driving advancements in various scientific disciplines.

One of the key factors contributing to English's preeminence in science is its ability to adapt and incorporate new terminology from diverse fields. The language's flexible nature has allowed it to absorb and integrate specialized vocabulary from fields such as medicine, genomics, and renewable technology, among others. This ongoing process of lexical expansion has enabled scientists to communicate complex ideas and share knowledge more effectively, fostering a culture of collaboration and innovation.

Moreover, English's linguistic features have been found to promote tentative and nuanced theorizing, which invites peer review and strengthens hypotheses. Studies have shown that English textual features like hedging, for instance, encourage scientists to express their ideas in a cautious and tentative manner, allowing for a more robust and rigorous testing of theories. This has contributed to the development of more reliable and accurate scientific knowledge, as well as a greater willingness to challenge and revise existing theories.

The inclusive nature of the English language has also played a significant role in promoting scientific progress. With its global reach and accessibility, English has facilitated the participation of diverse teams of scientists in international scientific discourse, fostering collaboration and driving innovation. This has helped to break down linguistic and cultural barriers, ensuring that scientific knowledge and discoveries are accessible to a wider audience, regardless of their geographical location or native language.

English's versatility, precision, and inclusive qualities have been instrumental in supporting scientific progress and the exchange of innovative discoveries over the centuries. Its ability to adapt and incorporate new terminology, promote tentative theorizing, and facilitate international collaboration has cemented its position as the primary language of science, fostering a culture of collaboration and driving advancements in various scientific disciplines.

Cross-Cultural Collaborations

Case studies showcasing how non-native English speakers contribute significantly to scientific innovation.

- Physicists Albert Einstein (German) and Satyendra Nath Bose (Indian): Pioneered quantum mechanics theories that reshaped physics in the 20th century. Published research together solely in English despite coming from different linguistic/cultural backgrounds.

- Biochemist Mahabir Gupta (Indian): Led teams developing typhoid vaccines at University of Maryland while training students internationally. Credits ability to conduct cross-border collaboration/knowledge sharing through English as a key factor in groundbreaking vaccine formulations.

- Neuroscientist Yukiko Goda (Japanese): Worked with labs across MIT and Harvard, co-authoring over 50 papers advancing understanding of synaptic function. Has said facility with English as lingua franca allowed her non-native perspectives to productively engage other thinking and jointly push frontiers.

- Chemist Younan Xia (Chinese): Known for integrating nanoscience and materials engineering innovations. Has raised millions in international funding/partnerships enabled through clear scientific communication styles cultivated in second language of English publications/presentations.

- Team led by Mikel Aizen (Spanish) at MARE Foundation: Conducting landmark ocean conservation/satellite tagging research with partners in over 20 countries. Demonstrates how collaborative project management and data interpretation occurs seamlessly across linguistic/cultural boundaries in English.

These case studies highlight English as an empowering equalizer accelerating scientific progress through inclusion of diverse global problem-solving capacities.

English: The Language of Science and Innovation ©

Chapter 5

Engineering Ingenuity

Engineering in the English Language

Discussing the inherent adaptability of English in expressing complex engineering concepts.

From its earliest forms, English has incorporated vocabulary from languages like Latin, Greek, and French to describe various technical domains like mathematics, mechanics, and architecture. Terms entered for elements, systems, measurements, and processes. Over time, the language finely tuned expressions for precise descriptions of designed structures and engineered works.

English grammar allows compounding technical nouns to efficiently conveys multi-faceted system components and properties in a single term. Its flexible style permits lengthy phrases integrating multiple specific modifiers before central nouns. This suits conveying the complex, interconnected nature of engineered artifacts and projects.

As engineering advanced, English proved adept atabsorbing newly coined terms. Neologisms flowed in from global innovators, facilitating universal sharing of developments. The language's adaptability means engineering terminology remains current thanks to fluid incorporation of applied research lexicons.

English also utilizes varied linguistic devices like nominalization favoring condensed, information-dense terms well-aligned with engineering's emphasis on parameterized systems. Its objectivity resonateswith quantitative, results-focused engineering problem-solving and reporting.

Centuries refining technical discourse endowed English with an ingrained engineering register. Its syntactic precisionand semantic range, coupled with ongoingterm evolution, make itparticularly equipped for the multidimensional communication demands of conceiving, implementing andstandardizing even the most sophisticated designs and breakthrough technologies internationally.

The centuries-long process of refining technical discourse has endowed English with a robust engineering register, characterized by syntactic precision and semantic range. This has enabled English to accommodate the multidimensional communication demands of conceiving, implementing, and standardizing complex designs and breakthrough technologies on a global scale.

The engineering register of English has evolved to include specialized vocabulary, idioms, and grammatical structures that facilitate the precise conveyance of technical information. This has been achieved through a combination of factors, including the language's inherent flexibility, the influx of technical terminology from various fields, and the ongoing process of term evolution.

English's syntactic precision allows for the creation of complex sentences that can express intricate technical concepts, while its semantic range enables the language to accommodate a broad array of technical terminology. This combination of features has made

English particularly well-suited for communicating the nuances of sophisticated designs and technologies.

Furthermore, the ongoing evolution of terms and phrases in English has helped to keep the language abreast of the latest technological advancements. New terms are constantly being coined to describe emerging technologies, and existing terms are adapted to fit the needs of new contexts. This fluidity has enabled English to maintain its relevance in the face of rapid technological change.

English's engineering register has been shaped by centuries of technical discourse, resulting in a language that is uniquely equipped to meet the multidimensional communication demands of the technological age. Its syntactic precision, semantic range, and ongoing term evolution have made it an indispensable tool for conceiving, implementing, and standardizing cutting-edge technologies on a global scale.

Global Engineering Networks

Highlighting how linguistic inclusivity facilitates global collaborations in engineering feats.

As the shared technical lingua franca, English has been instrumental in enabling massively complex international engineering projects that have furthered human development. Large-scale ventures like global satellite systems, particle accelerators and international space stations involve coordination of experts from dozens of countries.

Using English as a common work language allows seamless integration of diverse skills and cultural perspectives within multinational teams. Its worldwide intelligibility ensures all partners can equally contribute specialized knowledge toward problem-solving regardless of native tongue.

Project management conducted in English similarly provides an even linguistic platform for tasks like planning, risk assessment, procurement and reporting across stakeholders. Cultural translation costs are minimized while scope for innovative synergies is maximized.

English also underpins vast online engineering networks. Communities openly share blues using English, crowdfunding breakthrough concepts and crowd-testing prototypes. Emergent grassroots collaborations push what is possible in fields like renewable technologies, biomanufacturing and AI safety.

English acts as an inclusive linguistic equalizer that amplifies our species' capacity for collectively achieving feats greater than any single nation. By empowering truly global partnerships leveraging vast international talent pools, English continues propelling the engineering advances transforming society for the better.

Using English as a common language has numerous advantages in various fields, including engineering, education, recruitment, conferences, partnerships, and crisis coordination. Here are some specific benefits:

- Technical documentation and standards: English is widely used in technical documentation, recommendations, safety codes, and standards, ensuring interoperability and consistency across different countries and industries. This helps to avoid misunderstandings and miscommunications that could lead to accidents or errors.

- Education and training: English is the language of choice for massive open online courses (MOOCs), which educate global audiences and provide credentials for students worldwide. English textbooks standardize fundamental knowledge, making it easier for students to learn and apply concepts across different countries and cultures.

- Recruiting talent: Job openings in English reach a huge international talent pool, allowing companies to source qualified engineers from around the world, regardless of citizenship or location. This fosters a diverse and inclusive work environment, which can lead to innovative solutions and better decision-making.

- Conferences and workshops: English enables the transfer of best practices through global symposia, where experts can share knowledge and collaborate in real-time. Live collaborations at these events have led to breakthrough solutions and new technologies.

- Partnership diversity: Communicating in English helps bring more marginalized or minority perspectives to projects, fostering innovation through diverse cultural problem-solving approaches. This can lead to more creative and effective solutions, as well as a more inclusive and equitable work environment.

- Patent and research cooperation: Global intellectual property agreements in English encourage international research collaboration and the dissemination of new ideas. This leads to faster development and implementation of new technologies, benefiting society as a whole.

- Crisis coordination: English supported the mitigation of past disasters like Chernobyl or the 2011 Japan earthquake through coordinated international relief campaigns. In times of crisis, clear communication is crucial, and a common language can help ensure that aid is delivered efficiently and effectively.

English has become the language of engineering, innovation, and collaboration, providing numerous benefits across various fields. Its use as a common language facilitates the sharing of knowledge, ideas, and best practices, leading to better solutions, faster development, and a more inclusive and equitable work environment.

Chapter 6

The Language of Technology

Language as a Technological Enabler

Examining how English serves as a foundation for technological advancements.

From the earliest innovations, English embraced vocabulary to precisely describe new machinery, tools and devices. Over time, it refined nuanced terminology reflecting branching specializations within fields like computer science, robotics and biomedical engineering.

English grammar supports coining technical compounds representing multi-component systems and abstract software/network constructs. Its flexibility allows efficient communication of complex technological processes.

As a primary language of research journals and conferences, English facilitates rapid international dissemination of new technological concepts. Ubiquitous understanding accelerates refinement and applications.

Online in forums and documentation, English breaks down barriers for grassroots collaboration. Amateurs globally crowdsource solutions, propelling emergent technologies like AI assistants, prosthetics and renewable infrastructure.

As technology permeates all industries, English expands semantically to precisely articulate multi-disciplinary convergences. This fuels discovery at the frontier of previously unimagined integrated fields.

English's inherent qualities position it optimally to nurture the cross-pollination of ideas revolutionizing every sector. By serving as a shared medium for open partnership worldwide, English continues empowering unprecedented technological progress for the benefit of all humanity.

Additional points about how English serves as a foundation for technological advancement

Precise terminology: English absorbs neologisms to concisely yet accurately describe newly developed components, methodologies, coding paradigms, etc. This supports reproducibility.

Programming languages: Many popular coding languages like C++, Java, Python were designed/documented in English, leveraging its rigor for algorithmic logic and abstraction.

Patent systems: International intellectual property networks largely operate in English, incentivizing global dissemination and commercialization of innovative technologies.

Mass production standards: Common technical English allows coordination of international manufacturing and quality assurance standards essential for emerging sectors like renewable energy.

Investment/crowdfunding: Communicating concepts and targets in English helps startups attract international early backers to further develop their ideas.

Policy/ethics discussions: Debates around regulating emerging technologies from AI to bioprinting involve technical experts worldwide collaborating via shared English discourse.

Access to expertise: Technology companies leverage English to recruit top international talent that otherwise may not interact due to language/cultural barriers.

Technological Assimilation

How the language absorbs foreign ideas, accelerating technological growth.

English has long demonstrated an ability to seamlessly assimilate specialized terminology from other languages related to emerging technologies. This cross-pollination strengthens the language's descriptive power while accelerating dissemination of innovations globally.

For example, during the Renaissance, English incorporated scientific and engineering concepts from classical Greek and Arabic treatises, enriching early modern understandings. More recently, the language drew extensively on Chinese, Japanese and European sources amidst technology's explosive resurgence in the digital age.

Areas like electronics, robotics, aerospace and biotech witnessed waves of new terminology entering English due to this permeability. Words flowed in to precisely and succinctly represent foreign concepts, expanding the language's technological register.

Fluid adoption of specialized lexicons mirrors wider trends of globalization and cross-border fusion driving growth. Terminology assimilation integrates diverse cultural perspectives, fostering serendipitous connections and applications across fields. It also ensures universal spread and consolidation of state-of-the-art knowledge.

Over centuries, English emerged with an ingrained capacity to continuously remap its frontiers based on humanity's collective progress. This flexible response echoes the iterative nature of technology development itself and reinforces the language's role as society's indispensable platform for open exchange and progress.

English thrives through its inherent cultural and linguistic assimilation underpinning accelerating technological progress for the modern era and beyond. The language's fluid absorbency sustains harmony with our species' collaborative drive to constantly push boundaries.

How English absorbs technological ideas:

Neologisms: Foreign terms are efficiently anglicized through spelling/pronunciation adjustments while retaining core meaning. This streamlines communication.

Online dissemination: The internet multiplies opportunities for informal digital exposure to foreign technical lexicons, hastening assimilation into common English usage globally.

Academia: Research collaborations between international universities propagate terminology sharing through English publications, conferences, etc.

Private sector: Multinational tech/engineering firms provide venues for blended linguistic/cultural exposure that seeds lexical diffusion worldwide.

Patents: WIPO registers most intellectual property claims in English, systematically documenting new technologies in a single shared language.

Media: Popular science/trade press releases discussing foreign innovations introduce terminology to broad audiences in accessible English.

Immigration: Skilled migrant communities bridge linguistic/cultural understanding, enriching descriptive power of adopted host languages like English.

Online dictionaries: Resources like Wiktionary leverage crowdsourced efforts to continuously update English with cutting-edge foreign terms.

Chapter 7

The Cognitive Advantage

Cognitive Flexibility in English Speakers

Investigating studies that suggest linguistic influence on problem-solving and creativity.

Various experiments provide evidence that regular use of English may influence how people approach problems and foster creativity. Some key findings:

- Bilingualism research finds constantly switching between languages enhances cognitive control and idea generation abilities due to improved flexibility. As a global lingua franca, English exposes many to diverse perspectives.

- Studies using Raven's Progressive Matrices intelligence test report English speakers outperforming on items requiring analyzing relationships between changing patterns in novel ways. This hints at an advantage in flexibility and adaptability.

- Neuroimaging research indicates creative idea generation activites overlapping brain regions for processing syntactic complexity in

English sentences. Regular English use may strengthen these neural pathways.

- Analyses of interviews with high-achieving scientists found they frequently used hedging language and explored hypotheses tentatively when discussing work, similar to linguistic styles in English texts.

- Cross-cultural studies link collectivism values with preference for consistency while individualism associates with flexibility/openness to change. Pervasive individualism in English societies may influence related cognitive proclivities.

- Blended cultural/linguistic experiences abroad tend to correlate with gaining interdisciplinary perspectives and thinking "outside the box" - abilities English supports through permeable acquisition of varied specialty lexicons.

English engagement appears to stimulate neural mechanisms strengthening systemic and open-ended thinking - skills integral to creativity and innovative problem-solving. Further research can elucidate these cognitive advantages and linguistic influences.

Language and Thought Processes

Delving into theories exploring the correlation between language structure and cognitive abilities.

Ways language structure may shape thought processes:

Linguistic relativity hypothesis suggests grammar influences how we conceptualize the world. For example, English requires pronouns/order to convey agency, implying linear cause/effect thinking.

Studies find English speakers better at dissociating verb from object due to word order flexibility. This could enhance analytic, categorical reasoning over contextual/relational thinking.

Grammatical devices like nominalization pack complex ideas into single words, aligning with systematic/abstract conceptualization in fields like science/technology.

Weak grammatical constraints allow varied/tentative expression, possibly fostering open, flexible thinking styles while others favor consistency.

Languages with classifier systems partitioning semantic domains may facilitate expertise within categories vs cross-domain analogies.

Recursive/modular structures let English generate infinite expressions, mirroring brain's neural networking and ability to consider multiple perspectives.

Morphological complexity like inflection relays nuanced meanings/relationships, whereas analytic languages privilege disentangled referents/predicates.

Reading direction may influence association patterns, with left-to-right facilitating linear sequences for storytelling, experimentation.

While influences are debated, linguistic factors offer clues to how cognition naturally organizes based on inherent properties of the native language interface with environment/experiences. Further exploration is revealing relationships between language biases and proclivities.

The study of language and its relationship to cognition is a rapidly evolving field, and researchers are continuously uncovering new insights into how language influences our thinking and perception of the world. While the debate surrounding the influences of language on cognition is ongoing, linguistic factors offer valuable clues to how cognition organizes itself based on the inherent properties of the native language interface with the environment and experiences.

One of the key areas of research in this field is the study of language biases and proclivities, which refers to the ways in which language influences our thoughts, feelings, and behaviors. By examining the structural properties of language, such as grammar, syntax, and semantics, researchers can identify patterns and tendencies that shed light on how language shapes our cognition.

For example, studies have shown that the grammatical structures of a language can influence the way we think about objects and events. In languages that have a rich system of grammatical case markings, such as many Indo-European languages, the grammatical case of a noun can influence the way we perceive its relationship to other elements in a sentence. This can lead to differences in the way we categorize and conceive of objects and events in the world around us.

Another area of research involves the study of linguistic relativity, which is the idea that the language we speak influences the way we think and perceive the world. This concept has been the subject of much debate, with some researchers arguing that language has a profound influence on cognition, while others argue that the relationship is more complex and nuanced.

However, recent studies have provided evidence for the linguistic relativity hypothesis, suggesting that language does indeed play a role in shaping our cognition. For example, research has shown that speakers of different languages have different perceptions of color, with speakers of some languages having a greater distinction between certain colors than speakers of other languages. This suggests that language can influence our perception of the world around us, even in areas as fundamental as color perception.

While the debate surrounding the influences of language on cognition is ongoing, linguistic factors offer valuable clues to how cognition organizes itself based on the inherent properties of the native language interface with the environment and experiences. Further exploration into the relationships between language biases and proclivities is revealing new insights into how language influences our thinking and perception of the world, and is helping us to better understand the complex and dynamic relationship between language and cognition.

Relationship between language and thought:

Vocabulary size/domains: Languages with extensive lexicons for topics like spatial relations, colors or emotions may subtly guide related conceptual expertise.

Metaphors: Embedded metaphorical thinking in languages influences how abstract constructs are modeled and reasoned about.

Scripts/schemata: Different routines/actions encoded grammatically, like buying vs making, provide cognitive templates filter experiences.

Grammatical aspect: Languages with rich aspect may enhance awareness of circumstances/duration while others focus on conclusions.

Speech acts: Indirect/attention devices cross-linguistically signal nuances in intention/politeness that shape social cognition and cooperation.

Literacy: Learning to decode print impacts neural wiring for sequencing, sound-symbol awareness with ramifications for functions like memory.

Diglossia: Juggling multiple registers/dialects daily may confer adaptability cross-culturally through code-switching experience.

Language loss: Studies find forgetting first language associated with lost relational/categorization abilities requiring cultural re-learning.

Clearly the topic spans many factors and the relationship is bidirectional - language and thought continuously influence each other. More research unpacks this complex interface.

Chapter 8

Linguistic Adaptability

English as a Multifaceted Tool

Unveiling the adaptability of English to encompass a broad spectrum of ideas and concepts.

English is a versatile language that has evolved over time to accommodate a wide range of ideas and concepts. Its adaptability has enabled it to become a dominant language in various fields, including science, technology, business, and culture. Here are some ways in which English has demonstrated its adaptability:

- Vocabulary: English has a vast vocabulary that comprises words from various languages, including Latin, Greek, French, and many others. This has allowed it to absorb and adopt new words and expressions that reflect the changing times and advancements in various fields.

- Grammar: English grammar is flexible and allows for various sentence structures, which makes it possible to convey complex ideas and concepts. The language has also evolved to include new grammatical structures, such as the use of modal verbs and phrasal verbs, which have enabled it to express subtle nuances in meaning.

- Idioms and expressions: English is rich in idioms and expressions that convey complex ideas and emotions. These idioms and expressions are often culturally specific, which has helped English to adapt to different cultures and contexts.

- Pronunciation: English pronunciation has evolved over time, and different regions have developed their unique accents and pronunciations. This has enabled English to be spoken and understood by people from diverse backgrounds and regions.

- Spelling and orthography: English spelling and orthography have evolved to reflect the changing pronunciation and vocabulary of the language. This has made it easier for people to learn and use English, as well as to adapt it to their specific needs and contexts.

- Registers: English has multiple registers, including formal, informal, technical, and colloquial. This adaptability has enabled it to be used in various contexts, such as academic writing, casual conversation, technical writing, and professional communication.

- Acquisition: English is widely taught and learned as a second language, and its adaptability has made it easier for non-native speakers to acquire and use it effectively.

English is a multifaceted tool that has demonstrated remarkable adaptability over time. Its ability to absorb new words, expressions, and grammatical structures, as well as its flexibility in pronunciation, spelling, and register, has made it an indispensable language in various fields and contexts. Its adaptability has enabled it to become a dominant language in the world, and it continues to evolve and adapt to meet the changing needs and challenges of the global community.

Its ability to continually evolve and absorb new words, expressions and structures from other languages has been key to its widespread adoption and success on the global stage. A few thoughts in response:

- Flexibility and adaptability are indeed core strengths of English that have allowed it to spread so pervasively. By embracing linguistic change rather than resisting it, English remains a highly relevant lingua franca.

- This quality of absorbing outside influences without losing core elements of grammar and expression has enriched the language over time. It synthesizes ideas from diverse cultural sources into a shared vocabulary and system of communication.

- Areas like technology, science and business have benefited tremendously from English's evolving descriptive power. Its fluid adoption of specialized lexicons mirrors rapid advancement in these dynamic fields.

- Factors like spelling/pronunciation variability, register differences and room for creative expression give English an inclusive character. This likely enhances its appeal and utility across international communities with varied linguistic needs.

- Going forward, the challenges of ongoing globalization and digital transformation will continue testing English's flexibility. But its proven history of adaptation positions it well to facilitate international cooperation through periods of rapid sociocultural changes.

That adaptability has been English's defining strength, allowing it to serve an unparalleled role as a modern lingua mundi. Its inherent dynamic, synthetic qualities show no signs of limiting such influence into the future.

Embracing Foreign Elements

Analyzing how English effortlessly assimilates foreign terms and concepts.

English is a language that has always been open to embracing foreign elements, and it has a long history of assimilating terms and concepts from other languages. This ability to absorb and adapt foreign elements has been a key factor in English's success as a global language.

One of the main reasons why English is able to assimilate foreign terms and concepts so effortlessly is its vocabulary. English has a vast vocabulary that includes words from many different languages, including Latin, Greek, French, German, and many others. This means that English speakers have access to a wide range of words and expressions that can be used to describe complex ideas and concepts.

Another reason why English is able to assimilate foreign elements so easily is its grammar. English grammar is flexible and allows for a wide range of sentence structures, which makes it possible to express complex ideas and concepts in a variety of ways. This flexibility also makes it easier to incorporate foreign terms and concepts into English sentences.

Furthermore, English has a long history of borrowing words and expressions from other languages. This has been especially true during periods of colonialism and imperialism, when English was the language of power and influence. During these times, English speakers often borrowed words and expressions from the languages of the colonized peoples, which helped to enrich the English language and make it more expressive.

In addition, English has a strong tradition of linguistic hybridity. This means that English speakers are comfortable using words and expressions that are combinations of different languages. For example, the word "cosmopolitan" is a combination of the Greek word "kosmos" (meaning "world") and the Latin word "politanus" (meaning "citizen"). Similarly, the word "fusion" is a combination of the Latin word "fusio" (meaning "to mix") and the English word "fusion" (meaning "a combining of different things").

Finally, English has a strong tradition of neologism, which means that new words and expressions are constantly being created to describe new ideas and concepts. This means that English is always evolving and adapting to meet the needs of its speakers.

How English assimilates foreign elements:

- Permeability - English readily absorbs loanwords with minimal changes to spelling/pronunciation. This ease of incorporation streamlines communication of new ideas.

- Precise integration - Foreign terms are adapted contextually with associated technical or colloquial meanings intact. Nuances aren't lost in assimilation.

- Rapid propagation - Through globalization, the internet accelerates exposure and spread of loanwords among English speakers worldwide. Technology, pop culture terms spread especially fast.

- Enrichment of registers - Specialized lexicons from diverse fields continuously enhance English's descriptive range, from scientific to management terminology.

- Synthesis of perspectives - By integrating terminology from trade partners and former colonies, English synthesizes multicultural understanding and innovations.

- Flexible grammar - Loanwords can be easily modified through standard word formation rules like affixation, retaining the core foreign concept or object.

- Inclusiveness - Embracing outside lexicon makes English more universal and appeals to global communities, strengthening its role as a lingua franca.

- Living language -Fluid incorporation of neologisms ensures English remains ever-evolving and in sync with humanity's cross-cultural intellectual progress.

English's unique openness to foreign linguistic elements has been essential for maintaining relevance and increasing the language's unparalleled expressive scope in a globally interconnected world. It is a language that has always been open to embracing foreign elements, and it has a long history of assimilating terms and concepts from other languages. This ability to absorb and adapt foreign elements has been a key factor in English's success as a global language. Whether it is through vocabulary, grammar, borrowing, hybridity, or neologism, English continues to evolve and adapt to meet the needs of its speakers, and it will likely continue to do so for many years to come.

English: The Language of Science and Innovation ©

Chapter 9

Philosophical Inquiry in English

Philosophy and Linguistic Expression

Exploring how English fosters nuanced philosophical discussions and debates.

Philosophical inquiry in English has a rich history, with the language playing a crucial role in shaping and articulating complex philosophical ideas. English has been the primary language of philosophical discourse for several centuries, and it continues to be a vital tool for philosophers around the world. In this article, we will explore how English fosters nuanced philosophical discussions and debates, examining the ways in which the language's unique characteristics and expressions facilitate the exploration and communication of philosophical ideas.

- Vocabulary and Conceptual Frameworks

English has a vast vocabulary that includes an array of words and expressions that are specifically designed to convey complex philosophical concepts. This vastness of vocabulary allows philosophers to articulate subtle distinctions and nuances in their ideas, making it easier to convey sophisticated arguments and engage in productive debates. Moreover, English has a long history of borrowing words from other languages, which has enriched its philosophical vocabulary and enabled philosophers to draw upon a diverse range of conceptual frameworks.

- Grammar and Logical Structure

English grammar and sentence structure are designed to facilitate logical and clear expression of ideas. The language's syntax and punctuation allow philosophers to build complex sentences that convey intricate relationships between ideas, making it easier to express subtle arguments and logical connections. English also has a range of logical connectives, such as "therefore," "consequently," and "nevertheless," which enable philosophers to signal the logical relationships between their ideas and to construct persuasive arguments.

- Idiomatic Expressions and Figurative Language

English is replete with idiomatic expressions and figurative language that can be used to convey complex philosophical ideas in a nuanced and engaging way. Idioms, metaphors, and analogies enable philosophers to communicate abstract concepts in a more accessible and memorable way, making it easier for their audience to understand and engage with their ideas. Moreover, figurative language can be used to convey emotions and attitudes, which is essential in philosophical discussions that often involve questions of ethics, morality, and values.

- Clarity and Precision

English is known for its clarity and precision, which are essential qualities in philosophical writing and debate. The language's grammatical structure and vocabulary allow philosophers to express their ideas clearly and concisely, making it easier for their audience to follow their arguments and engage with their ideas. Moreover, English's emphasis on clarity and precision encourages philosophers to be rigorous in their thinking and to avoid ambiguity and vagueness in their arguments.

- Cultural and Historical Significance

English has played a significant role in the development of Western philosophy, particularly since the Enlightenment. The language has been the primary means of communication for many influential philosophers, such as John Locke, Immanuel Kant, and Bertrand Russell, among others. English has also been the language of many seminal philosophical texts, including John Stuart Mill's "On Liberty" and Jean-Paul Sartre's "Existentialism is a Humanism." This cultural and historical significance has endowed English with a rich philosophical heritage, making it a natural choice for philosophers around the world who seek to engage with the broader philosophical community.

- Global Reach and Accessibility

English has become the lingua franca of the world, with millions of people around the globe using it as a second language. This global reach and accessibility have made English an indispensable tool for philosophers who seek to engage with a broad audience and to participate in international philosophical discourse. English's global influence has also led to the development of diverse philosophical traditions and perspectives, fostering a rich and nuanced understanding of philosophical ideas and debates.

- Technology and Digital Communication

The rise of digital communication and technology has further enhanced English's role in philosophical inquiry. The internet and social media have made it easier for philosophers to share their ideas, engage with a wider audience, and participate in global philosophical discussions. Online platforms, such as blogs, podcasts, and video conferencing, have also facilitated the exchange of ideas and fostered collaboration among philosophers across different regions and time zones.

Ways English supports philosophical inquiry

Precision and clarity: English grammar emphasizes unambiguous subject-verb-object word order, facilitating precise articulation of abstract ideas, arguments and their logical relationships.

Nuance and hedging: Linguistic devices like hedges, modals and disclaimers encourage exploring multiple perspectives tentatively rather than definite claims, inviting critique and seeking truth collectively.

Flexibility: Compound words, hyphenation and long descriptive phrases allow expressing multifaceted concepts concisely without losing meaning, well-suited for complex philosophical theorizing.

Inclusiveness: As an international language, English provides a shared platform for exchanging diverse cultural lenses and worldviews, cross-pollinating philosophical traditions.

Established discourse: English philosophy literature's vast archives bequeath a richness of conceptual terminology, thought experiments and argumentation methodologies to build upon.

Living language: Thanks to ongoing lexical evolution, English naturally integrates neologisms synthesizing the latest interdisciplinary knowledge into philosophical inquiries.

Knowledge dissemination: English ensures universal accessibility of new philosophical schools of thought, their criticisms and communal refinement via academic journals, conferences and online debate.

English possesses linguistic qualities intrinsically empowering philosophical rigor, critical thinking, and collaborative truth-seeking on humanity's biggest questions. Furthermore, it has played a vital role in fostering nuanced philosophical discussions and debates. The language's unique characteristics, expressions, and cultural significance have made it an indispensable tool for philosophers around the world. English's voc

Language's Role in Ethical Contemplation

Discussing the influence of language on ethical frameworks and moral reasoning.

Language plays a crucial role in shaping our ethical frameworks and moral reasoning. The way we communicate and express our thoughts and ideas influences how we understand and approach ethical dilemmas. Here are some ways in which language impacts ethical contemplation:

- Conceptual frameworks: Language influences how we think about and categorize ethical concepts. For instance, the concept of "justice" is often associated with the idea of fairness, which is conveyed through words like "fair," "equal," and "impartial." Similarly, the concept of "compassion" is often linked to words like "empathy," "sympathy," and "kindness." The language we use to describe ethical concepts shapes our understanding of what they mean and how they relate to each other.

- Moral terminology: The words and phrases we use to describe moral concepts, such as "right," "wrong," "good," and "bad," influence how we evaluate ethical dilemmas. For example, the term "utilitarianism" describes a moral philosophy that prioritizes the greatest happiness for the greatest number of people. The language we use to describe ethical theories and principles can shape how we apply them in practice.

- Value-laden language: Language can be value-laden, conveying moral or ethical judgments. For instance, the phrase "human life is sacred" conveys a moral value that human life has inherent worth and should be protected. Similarly, words like "duty," "obligation," and "responsibility" convey a sense of moral obligation. The use of value-laden language can influence how we think about ethical issues and what we consider to be morally important.

- Emotive language: Language can also be emotive, evoking feelings and attitudes towards ethical issues. For example, the term "abortion" can evoke strong emotions and attitudes, depending on one's perspective on the issue. Emotive language can shape how we approach ethical dilemmas and influence our moral reasoning.

- Cultural influence: Language is shaped by cultural norms and values, which can influence how we think about ethical issues. For instance, some cultures prioritize collectivist values, such as loyalty to the group, while others prioritize individualist values, such as personal freedom. The language we use reflects these cultural values and can shape how we approach ethical dilemmas.

- Contextual factors: Language is also influenced by contextual factors, such as the setting, audience, and purpose of communication. For example, medical professionals may use technical language when discussing ethical issues related to patient care, while religious leaders may use spiritual language when discussing ethical issues related to their faith. The language we use is shaped by the context in which we use it, and this can affect how we approach ethical dilemmas.

- Linguistic ambiguity: Language can be ambiguous, with words and phrases having multiple meanings. For instance, the term "justice" can refer to both distributive justice (fair distribution of resources) and retributive justice (punishment for wrongdoing). Linguistic ambiguity can lead to misunderstandings and miscommunications, which can hinder effective ethical reasoning.

- Language evolution: Language is constantly evolving, with new words and phrases emerging to describe changing ethical concepts and issues. For example, the term "environmentalism" has emerged to describe a moral philosophy that prioritizes the protection of the natural world. The evolution of language reflects changes in societal values and ethical priorities.

- Translation and interpretation: Language can also influence how we approach ethical dilemmas across cultures and languages. The process of translation and interpretation can lead to misunderstandings and miscommunications, particularly when ethical concepts and values are culturally relative.

- Education and critical thinking: Finally, language plays a crucial role in ethical education and critical thinking. The development of critical thinking skills, such as analyzing arguments, evaluating evidence, and identifying biases, relies heavily on language proficiency. The language we use influences how we structure our thoughts, analyze ethical dilemmas, and communicate our ideas to others.

Language plays a vital role in shaping our ethical frameworks and moral reasoning. The way we communicate and express our thoughts and ideas influences how we understand and approach ethical dilemmas. By recognizing the influence of language on ethical contemplation, we can better navigate complex ethical issues and engage in more effective ethical reasoning.

Chapter 10

Innovation in a Global Context

Global Collaboration in English

Showcasing examples of groundbreaking collaborations transcending linguistic barriers.

Innovation in a global context is often characterized by collaboration and exchange across linguistic and cultural boundaries. English has emerged as a dominant language for global communication, facilitating collaboration and knowledge-sharing among diverse stakeholders. In this article, we will explore groundbreaking collaborations that transcend linguistic barriers, showcasing how English has enabled global innovation.

- International Space Station (ISS) Program

The ISS is a prime example of global collaboration, with astronauts and cosmonauts from over 15 countries working together to advance scientific knowledge and understanding. The program's success is largely attributed to the use of English as the common language for communication among the diverse crew members.

- CERN - The European Organization for Nuclear Research

CERN is a research organization that operates the Large Hadron Collider (LHC), the world's largest and most complex particle accelerator. Scientists and engineers from over 100 countries collaborate at CERN, using English as the primary language for communication. The LHC's discovery of the Higgs boson particle in 2012 is a testament to the power of global collaboration in scientific research.

- International Cancer Genome Consortium (ICGC)

The ICGC is a global collaboration of researchers, clinicians, and cancer centers working together to understand the genetic basis of cancer. With over 100 member countries, the ICGC relies on English as the common language for communication, enabling the sharing of data, research findings, and clinical practices.

- Global Health Initiatives

Global health initiatives, such as the Global Fund to Fight AIDS, Tuberculosis and Malaria, and the Bill & Melinda Gates Foundation, bring together experts from diverse countries and disciplines to tackle pressing global health challenges. English serves as the lingua franca for these initiatives, facilitating collaboration and knowledge-sharing among partners.

- Wikipedia - The Free Online Encyclopedia

Wikipedia is a prime example of global collaboration in the digital age. With over 50 million articles in over 300 languages, Wikipedia relies on English as the primary language for cross-cultural communication and knowledge-sharing. The platform's success is a testament to the power of collaborative knowledge-building and the role of English in facilitating global communication.

- International Business Collaborations

Multinational corporations and startups increasingly rely on English as a common language for communication and collaboration. Cross-border partnerships and joint ventures in industries such as technology, finance, and healthcare are commonplace, and English serves as the language of negotiation, innovation, and problem-solving.

- Global Education and Research Networks

The establishment of global education and research networks, such as the Global University Alliance and the Research Councils UK, has fostered collaboration among academics and researchers across borders. These networks promote knowledge-sharing, joint research projects, and student exchange programs, with English serving as the primary language for communication.

- International Artistic Collaborations

Artistic collaborations across languages and cultures have led to the creation of powerful works that transcend linguistic and cultural barriers. For instance, the global art project, "Inside Out," brought together artists from over 100 countries to create large-scale public installations that promote social and political change.

- Multilingualism and Translation Technology

The increasing availability of translation technology has enabled more accurate and efficient communication across languages. This has facilitated global collaboration in various fields, including business, education, and research. However, English remains a dominant language in many of these contexts, highlighting the ongoing importance of language skills in a globalized world.

- Global Challenges and Opportunities

The United Nations' Sustainable Development Goals (SDGs) represent a global commitment to address pressing challenges such as poverty, inequality, climate change, and social justice. Achieving these goals necessitates global collaboration, knowledge-sharing, and innovation, with English serving as a key language for communication and coordination.

English has emerged as a dominant language for global collaboration, facilitating communication and knowledge-sharing among diverse stakeholders. The examples discussed above demonstrate the critical role of English in driving innovation, scientific progress, and global problem-solving. However, the use of English should not be seen as a barrier to multilingualism or cultural diversity. Rather, it should be recognized as a powerful tool for fostering global understanding and collaboration, while promoting linguistic and

The use of English as a common language for global collaboration should not be seen as a barrier to cultural diversity or multilingualism. Below are some additional thoughts on this topic:

- While English facilitates communication across borders, it's important to recognize and respect other languages as crucial carriers of cultural heritage and knowledge. Major collaborations can and should find ways to incorporate local expertise and perspectives expressed in other tongues.

- More multilingual resources and translation services would help make big science projects and partnerships truly inclusive of diverse linguistic communities around the world. This could uncover new ideas and alleviate concerns about linguistic hegemony.

- Smaller, regional partnerships focused on local issues may find it preferable to conduct work primarily in shared local languages rather than English. This maintains cultural relevance.

- Bilingual and multilingual education should be encouraged to promote both English proficiency and the preservation of other languages. This nurtures global and local understanding simultaneously.

- International organizations leading major collaborations could adopt official multilingual policies and practices to foster a diversity of cultural and linguistic exchanges within the work.

- While English connects many, care should be taken not to assume fluency or conceptual equivalency across cultures even within collaborative contexts where it is used. Deeper understanding requires appreciation of linguistic and cultural nuances.

English enables vital collaborations but must do so in a way that celebrates multilingualism and cultural diversity, not replaces them. A both/and approach is best.

Language as a Catalyst for Global Innovation

Exploring how English acts as a facilitator for international innovation.

English plays a crucial but often overlooked role in driving global collaboration and progress. As the primary second language known by scientists, engineers, and professionals worldwide, it provides a common medium for sharing ideas and bringing diverse minds together in solving humanity's greatest challenges. When innovators from different countries can communicate directly in their work, cultural barriers break down and new perspectives emerge.

Through English, international partnerships form that would otherwise not be possible. Joint research projects are undertaken, integrating complementary areas of expertise from around the globe. Scientists publish findings not in silos but together in journals read internationally, accelerating our collective understanding. Students too are able to study abroad, exposing their home nations to outside thinking and transferring knowledge between cultures.

This cross-pollination spurs new breakthroughs as perspectives combine. An engineer in India reads about advances made by her peers in America, sparking fresh applications back home. A sociologist in Brazil joins economists from Europe to tackle social problems through an interdisciplinary lens. Through international workshops and conferences conducted in English, attendees are exposed to innovations presented by presenters worldwide, any of which could inspire their own problem-solving.

As a common technical language, English further ensures that global stakeholders speak the same innovation "language," whether drafting product specifications or filing international patents. Miscommunications pose less risk to collaborative projects. Standards developed jointly are clear to all. Emerging technologies can reach worldwide audiences fueled by grassroots digital communities that also use English as a bridge.

In this way, English acts as the essential connective tissue enabling globally distributed innovation networks to form and thrive. It spreads ideas far and wide, encouraging international minds to combine their problem-solving in ways that no single culture could achieve alone. In a world of vast and growing interconnectivity, English demonstrates how one shared tongue multiplies our collective progress for the benefit of all.

English: The Language of Science and Innovation ©

Chapter 11

Language, History, and Progress

Historical Perspectives on Linguistic Dominance

Examining how linguistic hegemony impacts the trajectory of human progress.

Language has been a vital tool for communication and expression throughout human history. However, the dominance of certain languages over others has had a profound impact on the trajectory of human progress. This essay will explore the historical perspectives on linguistic dominance and its effects on human progress.

The Rise of Linguistic Hegemony

Linguistic hegemony refers to the dominance of a particular language or languages over others, often accompanied by the marginalization or suppression of minority languages. The rise of linguistic hegemony can be traced back to the emergence of empires and colonization. As empires expanded their territories, they imposed their languages on the conquered populations, leading to the suppression of local languages and cultures.

One of the earliest examples of linguistic hegemony can be seen in the ancient empires of Egypt, China, and Rome. These empires imposed their languages on their territories, and the use of local languages was often discouraged or prohibited. The spread of Christianity in Europe during the Middle Ages also contributed to the rise of linguistic hegemony, as Latin became the language of the Church and the dominant language of scholarship and learning.

The Impact of Linguistic Hegemony on Human Progress

The impact of linguistic hegemony on human progress has been significant. On the one hand, the dominance of a particular language can facilitate communication and the exchange of ideas across different cultures and regions. This can lead to the spread of knowledge, innovation, and technological progress. For example, the spread of English as a global language has facilitated global communication and collaboration in fields such as science, technology, and business.

On the other hand, linguistic hegemony can also have negative consequences. The suppression of minority languages and cultures can lead to the loss of cultural diversity and the marginalization of minority communities. This can result in the exclusion of certain groups from access to education, economic opportunities, and political power. Moreover, the imposition of a dominant language can also lead to language loss, as speakers of minority languages may be forced to adopt the dominant language to avoid discrimination or marginalization.

Case Studies: English and Spanish

Two languages that have played a significant role in shaping the trajectory of human progress are English and Spanish. English has emerged as the dominant global language, with an estimated - 5 billion speakers worldwide. Its spread can be attributed to the rise of the British Empire in the 18th and 19th centuries, as well as the subsequent spread of American culture and technology in the 20th century.

Spanish, on the other hand, has played a significant role in shaping the history and culture of Latin America. The language was imposed on the indigenous populations of the Americas by Spanish conquistadors in the 16th century, leading to the suppression of many indigenous languages. Today, Spanish is the official language in many Latin American countries, and its influence can be seen in the culture, literature, and music of the region.

Linguistic hegemony has played a significant role in shaping the trajectory of human progress. The rise of dominant languages such as English and Spanish has facilitated communication and the exchange of ideas across different cultures and regions. However, the suppression of minority languages and cultures has also led to the loss of cultural diversity and the marginalization of minority communities.

It is essential to recognize the impact of linguistic hegemony on human progress and to strive for linguistic diversity and inclusivity. This can be achieved through the promotion of minority languages and cultures, as well as the recognition of the value of linguistic diversity in shaping the trajectory of human progress. By acknowledging the historical perspectives on linguistic dominance, we can work towards a more equitable and inclusive future for all languages and cultures.

Language's Influence on Cultural Narratives

Discussing how language shapes historical narratives and societal evolution.

Language plays a profound yet often underappreciated role in framing how societies understand themselves and their relationship to the past. The narratives a culture constructs are heavily dependent on the expressive ability and subtle nuances of the language they use to articulate their history, traditions and identity.

Different languages lend themselves particularly well to certain types of storytelling. For instance, rich verb conjugations in Romance languages facilitate relating histories through sequences of actions and events. Tonal languages allow conveying intricate layers of meaning and emotion when retelling folklore. Indigenous languages saturated with place-based vocabulary precisely locate cultural origins and migrations across landscapes.

Over generations, the narratives a language cultivates influence how a culture interprets its evolving values and social structures. Mythologies wrapped in poetic oral traditions cultivate particular religious worldviews. Historical epics recounted in writing solidify a national or ethnic sense of shared destiny. Folk tales inform cultural norms by examples of behaviors rewarded or punished within their fictional worlds.

When languages shift or are replaced due to colonization, technology or globalization, societies experience disruption as old narratives fade and new stories must be structured using foreign terms and styles of expression. Values associated with lost storylines may weaken while alternative ways of perceiving history and cultural continuity take hold.

Revitalizing endangered tongues reignites less familiar but equally valid narratives that challenge dominant trajectories and broaden incomplete perspectives. While linguistic diversity yields complexity, it also promotes cultural resilience and balanced understandings by presenting multitudes of lived experiences. Language profoundly determines the stories a people tell about themselves and what they pass down, for better and worse, across generations.

The structure and syntax of a language imprints on the style of narratives. For example, verb-final languages may lend themselves to more descriptive, place-based storytelling whereas subject-prominent languages facilitate more linear, plot-driven tales.

Figurative devices like metaphor, analogy and folk sayings embedded in a language color cultural storylines. They allow conveying complex ideas, lessons and traditions in memorable ways. Loss of these idioms impacts intergenerational transmission of history/values.

Oral narrative traditions vary greatly across languages and strongly mold a culture's storytelling conventions over time. Oral vs. written traditions prioritize different narrative perspectives and relationship to verifiable facts.

Censorship or promotion of approved narratives by colonizers often involved quashing indigenous languages and their opposing storylines to assert new history narratives. Two sides of the same history emerge.

Lexicon represents the conceptual categories of a culture. When loanwords enter a language from outside influence, they can reframe cultural storylines by enabling discussion of novel topics and shaping perception of change.

Languages define whose perspectives are centered - often marginalizing others. Revitalizing minority tongues helps broader understanding by bringing to light excluded narratives and untold histories from those cultures.

Storytelling conventions continue evolving with a language and societal changes. New genres may blend traditional schemata with modern influences, weaving innovative cultural narratives for the future.

The relationship between language and cultural narratives is deep and multifold. Understanding it remains crucial for appreciating a people's multidimensional history and identity

Language and cultural narratives are intimately connected, weaving together to form a complex system that reflects a people's history, beliefs, values, and identity. This relationship is multifaceted and profound, and understanding it is essential for appreciating the richness and diversity of cultures worldwide. In this article, we will delve into the intricate connection between language and cultural narratives, exploring how they shape and reflect each other.

- Language as a reflection of cultural narratives

Cultural narratives are the stories that a people tell about themselves, their history, and their place in the world. These stories are often passed down through generations, and they shape a people's identity, values, and beliefs. Language plays a crucial role in the transmission and preservation of these narratives, as it provides the means to express and communicate them.

The words, phrases, and expressions used in a language reflect the cultural narratives of a people. For instance, in many Indigenous cultures, there are no separate words for "he" and "she," instead using a single pronoun that encompasses both genders. This linguistic feature reflects the cultural narrative of gender equality and the importance of respecting all individuals, regardless of gender.

- Language as a tool for creating cultural narratives

Language is not only a reflection of cultural narratives but also a tool for creating them. Through language, people can express their experiences, thoughts, and emotions, weaving them into the fabric of their cultural narratives.

For example, the African American community has used language to create and preserve their cultural narratives, including the stories of slavery, segregation, and the struggle for civil rights. The use of African American Vernacular English (AAVE) has been instrumental in this process, as it has provided a means to communicate and express the community's unique history and identity.

- Language as a symbol of cultural identity

Language is a powerful symbol of cultural identity, as it embodies the values, beliefs, and experiences of a people. It is often used as a marker of belonging, and it plays a significant role in shaping a people's sense of self.

In many cultures, language is closely tied to the land and the ancestors who first spoke it. For instance, in many Indigenous cultures, the language is seen as a gift from the Creator, and it is deeply connected to the land and the community's spiritual beliefs.

- Language contact and cultural exchange

Language contact, or the interaction between speakers of different languages, has played a significant role in shaping cultural narratives. When people from different cultures interact, they often exchange words, ideas, and stories, leading to a rich and diverse cultural landscape.

For example, the English language has been shaped by numerous language contacts, including Latin, Greek, and French. This has resulted in a language that is filled with loanwords, calques, and other linguistic features that reflect the cultural exchange and borrowing that has occurred throughout history.

- Language and cultural revitalization

In recent years, there has been a growing interest in language revitalization, particularly among Indigenous communities. This involves reviving and revitalizing endangered languages, which are often seen as a critical part of cultural heritage and identity.

Language revitalization efforts often go hand-in-hand with cultural revitalization, as language is seen as a key component of a people's cultural narrative. By revitalizing their language, communities can also revitalize their cultural practices, beliefs, and traditions, ensuring their continued survival and transmission to future generations.

The relationship between language and cultural narratives is deep and multifold. Language reflects, creates, and shapes cultural narratives, while also serving as a symbol of cultural identity and a tool for cultural exchange and revitalization. Understanding this relationship is essential for appreciating the multidimensional history and identity of a people, and it remains crucial for promoting cultural understanding, respect, and preservation. By recognizing the significance of language in shaping our cultural narratives, we can better appreciate the richness and diversity of human experience.

English: The Language of Science and Innovation ©

Chapter 12

The Future of Linguistic Influence

Evolutionary Trajectory of English

Speculating on the future of English as a global language of intellectual pursuit.

As English continues spreading internationally, driven by digital connectivity and globalization, it will inevitably take on new hybridized forms. Regional varieties will become more prominent as societies creatively adapt English to their local linguistic cultures and conceptual needs.

We may see the emergence of new Englishes that splice in elements from other influential languages to efficiently discuss domains of expanding knowledge. Fields like science, technology and business will likely seed idioms that disseminate widely.

Simplified versions for cross-cultural business or technical standards may gain currency while artistic expression spawns many distinctive varieties. Digitally-facilitated spelling/grammar mutations will accelerate, unbound by traditional prescriptivism yet balancing fluid communicability.

In intellectual spheres, English may incorporate terminology from non-Western scholarly disciplines to reduce barriers to global participation and bringnew perspectives. Increased translation will also let other linguistic strengths inform interdisciplinary thought in English.

Pluralization is natural for a language interacting with many. Yet common understanding remains vital. Basic standardization combined with diversity may help English serve all while retaining fluid innovation. Recognition as a mosaic rather than a monolith supports continued spread as a primary lingua franca.

If guided judiciously yet organically, English could endure centuries more as the principal medium propelling worldwide exchange of science, culture and ideas. Its future success depends on sensitively weaving local nuance with universal intelligibility to foster truly inclusive and equitable global communication.

The future of English as a global language is likely to be shaped by a combination of technological advancements, globalization, and regional adaptations. As English continues to spread internationally, it will likely take on new hybridized forms, incorporating elements from other languages and cultures. This could lead to the emergence of new regional varieties of English, each with its own unique characteristics and idioms.

In fields such as science, technology, and business, English is likely to become more standardized and simplified, with a focus on cross-cultural communication and technical standards. However, in more creative fields such as art and literature, English is likely to spawn many distinctive varieties, with a focus on expression and style.

As English becomes more widespread, it is likely to incorporate terminology from non-Western scholarly disciplines, reducing barriers to global participation and bringing new perspectives to interdisciplinary thought. Increased translation will also allow other linguistic strengths to inform English, enriching its vocabulary and conceptual frameworks.

While English is likely to continue its dominance as a lingua franca, it is important to recognize its limitations and potential biases. Common understanding and standardization will remain vital, but they must be balanced with diversity and inclusivity. By embracing English as a mosaic rather than a monolith, we can foster continued spread and evolution while promoting truly equitable and global communication.

Ultimately, the future success of English as a global language depends on our ability to navigate the complex interplay between local nuance and universal intelligibility. By weaving together the diverse strands of language and culture, we can create a rich system of communication that is both dynamic and enduring.

Technological Impacts on Linguistic Evolution

Exploring how technology shapes the evolution and influence of languages.

The impact of technology on language evolution has been significant, and it continues to shape the way we communicate and interact with each other. Here are some ways in which technology has influenced language evolution:

- Globalization: The internet and social media have made it easier for people to connect with each other across the globe, leading to an increased exchange of ideas and cultural practices. This has resulted in the emergence of new linguistic patterns, such as the use of English as a global lingua franca, and the development of new dialects and hybrid languages.

- Language Contact: Technology has facilitated language contact between people who speak different languages, leading to the creation of new linguistic forms and the evolution of existing ones. For example, the use of machine translation and instant messaging apps has made it easier for people to communicate across language barriers, leading to the development of new lingual forms such as Spanglish and Chinglish.

- Language Change: Technology has also influenced language change by providing new ways of expressing ideas and emotions. For example, the use of emojis and emoticons has become a popular way of conveyancing emotions and ideas in digital communication, leading to a shift away from traditional linguistic forms.

- Language Death: Technology has also contributed to the death of certain languages, as people shift to more widely spoken languages or dialects that are more easily accessible through technology. For example, the use of indigenous languages in the Americas has decreased significantly due to the spread of European languages and the influence of colonialism.

- Language Revival: On the other hand, technology has also contributed to the revival of some languages that were previously considered endangered. For example, the use of digital tools and resources has made it easier for people to learn and speak indigenous languages, leading to a resurgence of interest in these languages.

- Language Standardization: Technology has also played a role in language standardization, as people are able to communicate with each other across different regions and countries. This has led to a greater uniformity in language usage and a decrease in regional dialects.

- Language Teaching: Technology has also made language teaching and learning more accessible and effective. For example, language learning apps and online courses have made it easier for people to learn new languages, and virtual reality technology has enabled people to practice language skills in a more immersive environment.

- Language Translation: Technology has also made language translation more accessible and accurate. Machine translation tools have become increasingly sophisticated, allowing people to communicate across language barriers with greater ease and accuracy.

Additional ways technology is shaping the evolution of languages

- The rise of online communication through tools like texts, emails and social media is influencing language use, especially among youth. Casual styles and abbreviations are becoming normalized.

- Machine translation is allowing more languages to access global information but may encourage code-switching between languages and the rise of "leetspeak". Over time, it could influence how some languages evolve or are preserved.

- Immersive technologies like virtual and augmented reality may spawn new technical or creative vocabularies as we describe emerging digital experiences. This could accelerate the development of new linguistic conventions.

- Automated writing through AI also poses risks, such as perpetuating societal biases, but may free humans to use language more creatively over time. It could spawn new genres or ways of thinking if used to augment rather than replace humans.

- The Internet acts as a meeting place for languages, letting obscure tongues find new speakers online. But it also increases endangerment risks to smaller languages through a "prestige effect" favoring global ones.

- Access to vast cultural works and translations online may help revitalize some endangered languages by inspiring new storytelling, but it also broadens foreign influences that could slowly change language identities.

- Technology both threatens linguistic diversity by amplifying dominant languages, while also supporting lesser-used ones. Its impacts will continuously reshape our cultural expression and thought processes in often unforeseen ways. Careful guidance can help maximize benefits and mitigate risks.

Technology has had a profound impact on the evolution of language, shaping the way we communicate and interact with each other. While it has contributed to the death of some languages, it has also revived interest in others and made language learning and teaching more accessible and effective.

Chapter 13

Linguistic Synthesis and Intellectual Hegemony

Amalgamation of Linguistic Legacies

Analyzing how English's incorporation of Latin, Greek, German, and French roots fortifies its capacity as a linguistic powerhouse.

The English language has emerged as a dominant force in the realm of global communication, and its linguistic synthesis has played a crucial role in shaping its trajectory. The amalgamation of various linguistic legacies, including Latin, Greek, German, and French, has contributed significantly to English's capacity as a linguistic powerhouse. This paper will examine the ways in which the incorporation of these roots has fortified English's position and influenced its evolution.

Latin Legacy

Latin, the language of the ancient Romans, has had a profound impact on the development of English. Latin's influence on English can be seen in the vocabulary, grammar, and syntax of the language. English has borrowed numerous words from Latin, especially in fields such as law, medicine, and science. Latin's grammatical structures, such as the genitive case and the subjunctive mood, have also been incorporated into English.

The influence of Latin on English can be seen in the many Latin-derived words that are used in everyday language. For example, words such as "hospital" (from the Latin "hospitale"), "doctor" (from the Latin "docere"), and "university" (from the Latin "universitas") all have Latin roots. Additionally, Latin's grammatical structures have shaped the way English speakers construct sentences. For instance, the use of the subjunctive mood in English, as seen in sentences such as "I suggest that he study harder," is a direct inheritance from Latin.

Greek Legacy

Greek, the language of ancient Greece, has also had a significant impact on the development of English. Greek has contributed to English's vocabulary, particularly in fields such as philosophy, science, and literature. Many English words have Greek roots, such as "philosophy" (from the Greek "philosophia"), "physics" (from the Greek "physikos"), and "literature" (from the Greek "litēratūra").

Furthermore, Greek's influence on English can be seen in the many borrowings from Greek mythology and literature. For example, words such as "fantasy" (from the Greek "phantasia"), "tragedy" (from the Greek "tragēdia"), and "comedy" (from the Greek "komodia") all have Greek roots. Additionally, Greek's grammatical structures, such as the use of the definite article, have been incorporated into English.

Germanic Legacy

Germanic languages, such as Old English and Old Norse, have also played a significant role in shaping the English language. Germanic languages have contributed to English's vocabulary, particularly in fields such as warfare, agriculture, and daily life. Many English words have Germanic roots, such as "father" (from the Old English "fæder"), "mother" (from the Old English "mōdor"), and "house" (from the Old English "hūs").

Furthermore, Germanic languages have influenced English's grammatical structures, such as the use of modal verbs and the word order in sentences. For example, the use of the modal verb "will" (from the Old English "willa") to express future tense is a direct inheritance from Germanic languages. Additionally, the word order in English sentences, such as the subject-verb-object (SVO) word order, is similar to that of Germanic languages.

French Legacy

French, the language of France, has had a significant impact on the development of English, particularly in the Norman Conquest of 106- French has contributed to English's vocabulary, particularly in fields such as law, government, and cuisine. Many English words have French roots, such as "government" (from the French "gouvernement"), "law" (from the French "loi"), and "restaurant" (from the French "restaurant").

Furthermore, French has influenced English's grammatical structures, such as the use of the definite article and the formation of sentences. For example, the use of the definite article "the" (from the French "le") to indicate a specific noun is a direct inheritance from French. Additionally, the use of the subjunctive mood in

English, as seen in sentences such as "I suggest that he study harder," is similar to the use of the subjunctive mood in French. The incorporation of Latin, Greek, Germanic, and French roots has fortified English's

On how English's synthesis of linguistic roots fortified its intellectual influence

- Adopting technical, legal and scientific vocabulary from Latin and Greek equipped early modern English with precise terminology needed to discuss burgeoning fields. This stimulated advances.

- Absorbing Romance-language words via Norman French infused English with a diversity of concepts from continental philosophical and literary traditions.

- Integrating Germanic influences retained linguistic ties to English's cultural origins while also gaining perspectives from Central European thought.

- Weaving vocabularies from conquering and colonized languages afforded English descriptions of knowledge from societies worldwide, broadening its semantic range.

- Compound words allowed unpacking meanings of complex loanwords while streamlining communications, promoting exchange of ideas.

- Through education and empire, English's lexical synthesis disseminated countless interdisciplinary innovations, attracting 更多使用者.

- Its linguistic mosaic reinforced English as the preeminent medium of global intellectual, political and commercial discourse and publications.

By skillfully amalgamating influences over centuries, English organically constructed an unparalleled linguistic instrument driving Western hegemony through ideas as much as force of arms. This amassed legacy persisted as its crucial competitive advantage.

Dominance of Enlightened Languages

Tracing the historical evolution of intellectual enlightenment across ancient Greek, Roman, Renaissance, and German thought and their amalgamation in English.

The concept of intellectual enlightenment has been a driving force behind the development of various languages throughout history. Each language has contributed significantly to the evolution of thought and has played a crucial role in shaping the dominant languages of today. In this essay, we will explore the historical evolution of intellectual enlightenment across ancient Greek, Roman, Renaissance, and German thought and trace their amalgamation in English.

- Ancient Greek Thought

Ancient Greek thought laid the foundation for Western philosophy and intellectual enlightenment. The Greeks believed in the power of reason and critical thinking, which led to the development of various philosophical schools, such as Platonism and

Aristotelianism. The Greek language was the medium through which these ideas were expressed, and it played a crucial role in shaping the intellectual landscape of the time.

The ancient Greeks also made significant contributions to the fields of science, mathematics, and literature. The works of Homer, Aristotle, and Plato, among others, have had a lasting impact on Western thought. The Greek language's emphasis on reason, logic, and clarity has influenced the development of many modern languages, including English.

- Roman Thought

Roman thought built upon the foundations laid by the ancient Greeks. The Romans adopted many Greek ideas and integrated them into their own culture. Latin, the language of the Romans, became the lingua franca of the Roman Empire and played a central role in the spread of intellectual enlightenment.

The Romans made significant contributions to the fields of law, governance, and engineering. Their legal system, for example, has had a lasting impact on modern Western legal systems. The Latin language's emphasis on precision and clarity has influenced the development of many modern languages, including English.

- Renaissance Thought

The Renaissance was a time of intellectual rebirth in Europe, marked by a renewed interest in classical Greek and Roman thought. The period saw the emergence of humanism, which emphasized the potential of human beings to achieve intellectual and artistic excellence. The Renaissance also saw the rise of the scientific method, which emphasized empirical observation and experimentation.

The Renaissance was a time of great linguistic innovation, with the emergence of new languages such as Italian, French, and Spanish. These languages were influenced by the classical languages of Greek and Latin and played a significant role in shaping the intellectual landscape of Europe.

- German Thought

German thought has had a profound impact on modern intellectual enlightenment. The German language has played a central role in the development of philosophy, science, and literature. The works of Kant, Hegel, and Nietzsche, among others, have had a lasting impact on Western thought.

The German language's emphasis on precision and clarity has influenced the development of many modern languages, including English. The German tradition of philosophical thought has also had a significant impact on the development of modern intellectual enlightenment.

- Amalgamation in English

English has emerged as the dominant language of modern intellectual enlightenment. The language has been shaped by the influences of ancient Greek, Roman, Renaissance, and German thought. English has incorporated words and ideas from these languages, and its grammar and syntax have been influenced by their linguistic structures.

The emphasis on reason, logic, and clarity that characterizes English has its roots in the classical languages of Greek and Latin. The language's ability to absorb words and ideas from other languages has allowed it to become a lingua franca, capable of expressing complex intellectual ideas.

The development of intellectual enlightenment has been shaped by the evolution of various languages throughout history. Each language has contributed significantly to the evolution of thought, and their amalgamation in English has created a rich and diverse linguistic landscape. The emphasis on reason, logic, and clarity that characterizes English has its roots in the classical languages of Greek and Latin and has been influenced by the linguistic and intellectual traditions of Renaissance and German thought.

The dominance of English as a language of intellectual enlightenment is a reflection of the rich cultural heritage that it represents. The language's ability to absorb and integrate ideas from other languages and cultures has made it a powerful tool for the expression of complex intellectual ideas. As the world becomes increasingly interconnected, the importance of English as a language of intellectual enlightenment will only continue to grow.

Analysis of how intellectual enlightenment evolved across different languages and cultures, and ultimately amalgamated in English

Ancient Greek laid philosophical foundations through nuanced terminology developed by Socrates, Plato, and Aristotle to debate logic, rhetoric, politics and natural sciences.

Latin disseminated Greek traditions throughout the Roman Empire, established as the scholarly language of the medieval world through works of Cicero, Ovid, Virgil.

Renaissance humanism saw a rebirth of interest in Classical Greek/Latin texts. Vocabulary from these languages infused English and other modem European languages.

German philosophy from Kant, Hegel focused enlightenment onto subjective human understanding/experience. Their terminology entered English via translations.

18th century French rationalism from philosophers like Descartes, Voltaire popularized analytical styles of critique that shaped Enlightenment thinking.

Emergent English periodicals/societies in the 17th-18th centuries amalgamated all these philosophical streams into a new intellectual movement through lexical borrowing.

English's flexible grammar structures and openness to loanwords seamlessly integrated diverse strands of thought, consolidating ideas across Europe and Americas.

So while each era contributed to human progress, English emerged uniquely positioned through its history of linguistic synthesis to consolidate the fruits of global intellectual enlightenment into a shared discourse.

English: The Language of Science and Innovation ©

Chapter 14

Morphological Prowess of English

Morphological Diversity and Cognitive Ease

Exploring how the amalgamation of linguistic elements within English enhances cognitive fluency and conceptual assimilation.

English's rich and diverse morphological processes have conferred cognitive advantages for its speakers by facilitating efficient conceptual expression and assimilation. Its ability to generate new words through derivation, compounding, blending and inflection mirrors humans' natural tendencies towards conceptual blending and thinking in analogical patterns.

Deriving words from other languages or forming new ones out of existing lexical items allows English to succinctly unpack the meanings of technical, unfamiliar or multifacted concepts into single terms. This boosts retention and understanding of specialized domains. Compounding in particular reflects how the mind combines discrete notions into more complex, abstract ideas. Blending creatively combines elements of different words in a playful yet analytically rigorous manner, stimulating lateral thinking.

English further benefits cognition through its use of inflectional contrasts that reinforce recognition of variances in form-meaning relationships. Its enormous lexicon moreover lets individuals precisely contextualize meanings through nuanced synonym choices, optimizing effective discourse. Additionally, the obscured etymologies of many English words reduce dependence on their language of origin for conceptual processing.

English morphology has evolved to closely align with humans' natural cognitive proclivities for economical representation, pattern-finding and category-crossing in knowledge assimilation. Its diverse word formation processes enhance processing fluency and conceptual expression. This adaptive congruence with underlying thought mechanisms reinforces English's optimal role as the preeminent vessel of global ideas exchange.

English's diverse morphological processes indeed stand as a testament to its adaptability and resonance with the cognitive proclivities inherent in human thought.

The derivational prowess of English, drawn from various linguistic roots, serves as a cognitive amplifier. By encapsulating multifaceted or technical concepts within concise terms, it becomes a vehicle for efficient comprehension and retention. Compounding, a testament to the mind's ability to synthesize discrete notions into complex ideas, reflects not only linguistic complexity but also cognitive agility and abstraction.

Moreover, the playful yet analytically rigorous nature of blending mirrors the brain's capacity for lateral thinking and creative synthesis. Through inflectional contrasts, English reinforces the recognition of nuanced form-meaning relationships, aligning with the human tendency to discern subtle variations in concepts.

The expansive lexicon of English empowers individuals to precisely contextualize ideas by employing nuanced synonyms, optimizing communication and discourse. The obscured etymologies of many English words further liberate conceptual processing from strict dependencies on their original languages, fostering a more universal comprehension.

English's evolution in morphology remarkably aligns with the intrinsic cognitive tendencies of humanity, facilitating economic representation, pattern recognition, and cross-categorical assimilation of knowledge. This congruence fosters processing fluency and nuanced expression, reinforcing English's role as the preeminent conduit for global exchange of ideas and knowledge.

Cultural Infusion and Language Evolution

Discussing how the infusion of diverse linguistic elements enriches English and fosters a fertile ground for intellectual exploration.

English has proven uniquely capable of organic cultural infusion and continual evolution, rendering it a fertile medium for intellectual growth. The incorporation of vocabulary from the languages it encountered has repeatedly enriched its expressive breadth.

As English spread throughout the world, it absorbed words from myriad cultures, synthesizing diverse perspectives into a shared semantic resource. Terms from colonial ties and global tradeinfused new technical, social and artistic concepts. This preserved cross-pollination of ideas during periods of exploration.

Linguistic exchange went reciprocal as well, with words like barbecue, hoopla and hodgepodge entering influence languages. Such balanced assimilation illuminated shared experiences while respecting cultural uniqueness.

Online connectivity now exponentially amplifies exposure to localized terminology worldwide. Neologisms rapidly disseminate into common parlance through Internet memes and direct sharing on platforms. This stimulates semantic evolution matching humanity's interconnectivity.

English's absorbency and flexibility create an perpetually expanding lexical repository. Its fluid morphology moreover derives new meaning representations that maintain precision amid change. Together, these qualities ensure the language remains an optimally rich medium for a diversity of intellectual pursuits globally. English's history of dynamic cultural incorporation distinguishes it as the preeminent vessel catalyzing our species' collective learning.

The English language has a remarkable ability to adopt and incorporate words from other languages, which has helped it become a dominant language in the world. This ability has allowed English to absorb new ideas and concepts, and to evolve in response to changing cultural and technological contexts.

One example of this is the way that English has incorporated words from colonial languages, such as French and Spanish, into its vocabulary. This has helped to create a shared cultural identity and facilitate communication between people from different parts of the world.

Another example is the way that English has incorporated words from other languages, such as Chinese and Arabic, into its vocabulary. This has helped to create a more diverse and inclusive language, and has allowed English speakers to communicate more effectively with people from different cultural backgrounds.

In addition to its ability to incorporate words from other languages, English has also developed a rich and diverse vocabulary through its own internal processes. This has allowed English to express complex ideas and concepts in a way that is unique and powerful.

The ability of English to incorporate words from other languages and to evolve in response to changing cultural and technological contexts has helped to make it a dominant language in the world. Its flexibility and adaptability have made it a valuable tool for communication and a reflection of the diverse cultures and experiences of its speakers.

The dynamic nature of English as a language is deeply intertwined with its global dominance. Its remarkable capacity to assimilate words from diverse languages, embracing cultural nuances and technological advancements, has propelled its ascendancy as a lingua franca.

English's openness to adopt terms from various languages speaks volumes about its adaptability and inclusivity. By incorporating words from different cultures, it has become a mosaic reflecting the rich history of human experiences. This linguistic flexibility allows English to evolve organically, mirroring the ever-changing landscape of our interconnected world.

Moreover, English's responsiveness to cultural and technological shifts endows it with a contemporary relevance that transcends geographical borders. As cultures interact and technologies advance, English serves as a conduit, effortlessly integrating new concepts and ideas into its lexicon. This adaptability not only enriches the language but also facilitates smoother communication among diverse communities, fostering mutual understanding and collaboration.

Furthermore, the dominance of English in various domains, including academia, business, science, and entertainment, stems from its ability to evolve and cater to the evolving needs of its speakers. Its flexibility enables it to articulate complex thoughts, facilitate global trade, disseminate knowledge, and serve as a platform for cultural expression on a worldwide scale.

The resilience and adaptability of English have propelled it to a position of global significance. Its ability to embrace linguistic diversity and reflect the ever-evolving human experience underscores its pivotal role as a powerful instrument for communication, connection, and cultural exchange in our interconnected world.

Chapter 15

Cognitive Resonance in Linguistic Synthesis

Cognitive Comfort in Linguistic Syncretism

Investigating why the human brain finds cognitive resonance in English due to its fusion of familiar linguistic elements.

Linguistic synthesis, the process of combining familiar linguistic elements to create new words, phrases, and languages, has been a crucial aspect of human communication throughout history. This process has resulted in the development of various languages, dialects, and hybrid languages, each with its unique set of linguistic features. English, in particular, has emerged as a prime example of a language that has undergone extensive linguistic synthesis, with its vocabulary comprising a rich blend of words from various languages. This article aims to explore the cognitive aspects of linguistic synthesis in English and investigate why the human brain finds cognitive resonance in this language.

Cognitive Resonance:

Cognitive resonance refers to the mental comfort and familiarity that individuals experience when they encounter familiar linguistic elements. This phenomenon is not limited to language alone but can be observed in various aspects of human cognition, such as memory, attention, and perception. In the context of language, cognitive resonance occurs when listeners or readers recognize and process familiar words, phrases, or grammatical structures.

Linguistic Syncretism:

Linguistic syncretism is the fusion of familiar linguistic elements to create new words, phrases, or languages. This process is a fundamental aspect of language development, and it has contributed significantly to the evolution of English. Linguistic syncretism in English has resulted in the creation of new words, such as "smog" (a blend of "smoke" and "fog"), "brunch" (a combination of "breakfast" and "lunch"), and "motel" (a fusion of "motor" and "hotel").

Cognitive Comfort in Linguistic Syncretism:

The human brain finds cognitive comfort in linguistic syncretism due to the familiarity of the linguistic elements involved. When listeners or readers encounter a new word that is a blend of familiar elements, they can quickly recognize and process it, as it leverages their existing knowledge and mental representations. This recognition and processing facilitate mental comfort, as the brain can efficiently integrate the new word into its linguistic repository.

Additionally, linguistic syncretism in English has also led to the creation of new meanings and concepts, which have contributed to the language's expressive power. The fusion of familiar linguistic elements has enabled the language to convey complex ideas and emotions more effectively, making it a rich and versatile medium of communication.

Neural Basis of Cognitive Resonance:

The neural basis of cognitive resonance in linguistic syncretism can be attributed to the way the brain processes and stores linguistic information. Research in neuroscience has shown that language processing involves a network of brain regions, including the left hemisphere of the inferior frontal gyrus (Broca's area), the left posterior inferior temporal gyrus (Wernicke's area), and the anterior cingulate cortex. These regions are responsible for various aspects of language processing, such as syntax, semantics, and phonology.

When listeners or readers encounter familiar linguistic elements, the brain can activate these regions more efficiently, as the information is already stored in long-term memory. This activation leads to a sense of cognitive comfort and familiarity, as the brain can quickly retrieve and integrate the relevant information.

Cognitive resonance in linguistic syncretism is a significant aspect of language processing and communication. The human brain finds cognitive comfort in English due to its fusion of familiar linguistic elements, which leverages the brain's existing knowledge and mental representations. This recognition and processing facilitate mental comfort, as the brain can efficiently integrate new words and meanings into its linguistic repository. The neural basis of cognitive resonance can be attributed to the way the brain processes and stores linguistic information, with the activation of brain regions such as Broca's and Wernicke's areas playing a crucial role in language processing.

Consequently, linguistic syncretism in English has contributed significantly to the language's expressive power and versatility, making it a prime example of a language that has undergone extensive linguistic synthesis. The cognitive aspects of linguistic syncretism highlight the importance of familiarity and recognition in language processing and communication, and offer insights into the neural mechanisms that underlie language comprehension and production.

The human mind appears well-adapted for cognitive comfort from English's fusionary nature stemming from four factors:

- Familiarity - English incorporates familiar lexicons from former regional lingua francas like Latin, Norse, French that retain remnants across modern European tongues. This promotes comprehension through mnemonic associations.

- Pattern recognition - English word formation follows recognizable morphological conventions like compounding and derivation absorbed from precursor languages. Its analytic grammar employs subject-verb-object syntax paralleled across sister Germanic languages. These formulaic traits ease parsing.

- Conceptual blending - Merging components from diverse lexicons mirrors the brain's penchant for combining discrete concepts nonlinearly into novel assemblages through analogy-making and metaphor. This taps into inherent cognitive mechanisms.

- Adaptive recursion - Just as the human mind accommodates new knowledge by assimilating it into preexisting neural networks, English fluidly integrates foreign terminology into its standing morphological/grammatical infrastructure. This maintaining structural cohesion mitigates processing demands.

Together, these factors suggest the cognitively congruent nature of English stems from its continued fusion of elements bearing ancestral linguistic familiarity. Its syntactical assimilation of outside ideas aligns aptly with intrinsic tendencies in human thought architecture.

Enhanced Idea Expression and Comprehension

Illustrating how the amalgamation of Latin, Greek, German, and French elements in English amplifies the expression and comprehension of ideas.

The English language, with its diverse blend of Latin, Greek, German, and French elements, offers a unique advantage in expressing and comprehending complex ideas. This amalgamation of linguistic roots enables English to convey a wide range of meanings and nuances, making it a versatile tool for communication.

- Latin: Latin, the language of the ancient Romans, has had a profound impact on English. Many Latin words have been incorporated into English, especially in fields such as law, medicine, and science. Latin roots allow English speakers to convey complex ideas with precision and clarity. For example, the Latin root "tele-" means "far" or "distant," and is found in words like "telephone," "television," and "telecommunications." This shared root facilitates understanding across languages and cultures, enhancing the expression and comprehension of ideas.

- Greek: Greek, the language of ancient Greece, has also had a significant influence on English. Many Greek words have been incorporated into English, particularly in fields such as philosophy, mathematics, and literature. Greek roots provide English speakers with a rich vocabulary for expressing complex ideas and concepts. For example, the Greek root "logy-" means "study of," and is found in words like "biology," "psychology," and "theology." This shared root enables English speakers to communicate complex ideas with greater precision and clarity.

- German: German, the language of Germany, has also played a role in shaping the English language. Many German words have been incorporated into English, particularly in fields such as engineering, technology, and business. German roots provide English speakers with a robust vocabulary for expressing complex ideas and concepts. For example, the German root "werk-" means "work," and is found in words like "workshop," "factory," and "machine." This shared root facilitates understanding across languages and cultures, enhancing the expression and comprehension of ideas.

- French: French, the language of France, has had a significant impact on English, particularly in fields such as cuisine, fashion, and art. Many French words have been incorporated into English, providing speakers with a rich vocabulary for expressing complex ideas and concepts. For example, the French root "chateau-" means "castle," and is found in words like "chateau," "chateau-briand," and "chateau-style." This shared root enables English speakers to communicate complex ideas with greater precision and clarity.

Tthe amalgamation of Latin, Greek, German, and French elements in English amplifies the expression and comprehension of ideas. The shared roots and vocabulary across languages enable English speakers to communicate complex ideas with greater precision and clarity, facilitating understanding across cultures and languages. This unique blend of linguistic elements makes English a versatile and powerful tool for communication, allowing speakers to express and comprehend complex ideas with greater facility than ever before.

The syncretic nature of English has significantly enhanced both the expression and comprehension of ideas through:

Expression:

- Precision - Technical, scientific and academic vocabulary from Latin and Greek allows finely detailed conceptual articulation.

- Nuance - More emotive and metaphorical terms from French and Latin convey subtlety and layered meanings.

- Description - Anglo-Saxon Germanic vocabulary provides a robust base for narrating phenomena and processes simply.

- Neologism - The open architecture allows generating novel terminology capturing emerging notions.

Comprehension:

- Familiarity - Shared Indo-European roots with other languages improve decoding unfamiliar contexts.

- Etymology - Tracing a word's origins across assimilated languages improves retention of relating concepts.

- Associations - Remnant connections to precursor lexicons activate supplementary neural pathways.

- Synergies - Overlaps in combinatorial rules like compounding/derivation capitalize on preexisting linguistic intuitions.

- Connectivity - Words interface with a vast web of related semantics enhancing inferencing of unstated relations.

Together, English's synthesized architecture maximizes both succinct expression and enrichment of implicit comprehension compared to its component languages in isolation. This catalyzes dissemination and communal refinement of knowledge.

Indeed, the synthesized architecture of English acts as a catalyst for both concise expression and enriched comprehension, surpassing its individual component languages in isolation. This unique amalgamation facilitates the dissemination and collective enhancement of knowledge within the global community.

English's amalgamated structure, drawing from diverse linguistic roots, enables succinct expression by encapsulating multifaceted concepts into single, precise terms. This linguistic efficiency optimizes communication, allowing for the swift transmission of ideas across cultural and geographical boundaries.

Moreover, the synthesized nature of English fosters implicit comprehension, surpassing the capabilities of its component languages when considered individually. The fusion of various linguistic elements within English enriches its lexicon, enabling nuanced interpretations and a depth of meaning that transcends the limitations of its original sources.

This fusion of linguistic influences not only aids in the efficient transmission of knowledge but also encourages communal refinement. Through continuous interactions and exchanges, the language evolves and adapts, accommodating new ideas and perspectives. This iterative process of refinement occurs organically within the English-speaking community, enriching the language's depth and breadth of expression.

Ultimately, English's synthesized architecture serves as a vehicle for the rapid exchange and communal advancement of knowledge, leveraging both concise expression and enriched comprehension. It stands as a testament to the power of linguistic synthesis in fostering a global discourse that transcends boundaries and cultivates a shared reservoir of collective wisdom and understanding.

Chapter 16

Language as a Catalyst for Universal Understanding

Universalizing Linguistic Constructs

Exploring how the assimilation of diverse linguistic elements in English fosters a universal platform for idea dissemination and comprehension.

The English language has evolved over time to become a melting pot of diverse linguistic elements from various cultures and languages. This assimilation has not only enriched the language but also created a universal platform for idea dissemination and comprehension. The incorporation of words, phrases, and grammatical structures from different languages has enabled English to become a lingua franca, facilitating communication and understanding among people from diverse backgrounds.

One of the primary ways in which English has universalized linguistic constructs is through the adoption of foreign words. Words from languages such as Latin, Greek, French, and German have been incorporated into English, providing a common vocabulary for people to communicate complex ideas. For instance, the word "telephone" is derived from the Greek words "tele" (meaning "far") and "phone" (meaning "voice"), while the word "computer" is derived from the Latin word "computare" (meaning "to calculate"). These words have become an integral part of the English language, allowing people to communicate ideas and concepts across linguistic and cultural boundaries.

Another way in which English has universalized linguistic constructs is through the adoption of loanwords. Loanwords are words that are borrowed from another language and incorporated into a different language. English has borrowed words from various languages, such as Hindi, Chinese, and Arabic, to describe concepts and objects that are specific to those cultures. For example, the word "curry" is derived from the Hindi word "kari," meaning "sauce," while the word "sushi" is derived from the Japanese word "寿司" (sushi), meaning "vinegared rice." The incorporation of loanwords has enriched the English language, allowing people to communicate ideas and concepts that are specific to different cultures.

In addition to the adoption of foreign words and loanwords, English has also universalized linguistic constructs through the use of calques. Calques are words or phrases that are borrowed from another language and translated into English. For example, the phrase "joie de vivre" is a French phrase that has been calqued into English as "joy of living." The use of calques has enabled English speakers to communicate ideas and concepts that are specific to other cultures, while also maintaining the nuances and connotations of the original language.

Furthermore, the assimilation of diverse linguistic elements in English has also led to the development of new grammatical structures. For instance, the use of the auxiliary verb "to be" in the present tense is a feature that has been borrowed from other languages, such as French and German. This has enabled English speakers to communicate complex ideas and concepts with greater precision and clarity.

The universalizing of linguistic constructs in English has also facilitated the development of new words and phrases that are specific to the language. For example, the word "selfie" was recently added to the Oxford English Dictionary, and it refers to a photograph taken of oneself. This word has become widely used across the globe, illustrating how English has become a platform for universal communication and understanding.

The assimilation of diverse linguistic elements in English has universalized linguistic constructs, creating a platform for idea dissemination and comprehension that transcends cultural and linguistic boundaries. The adoption of foreign words, loanwords, calques, and new grammatical structures has enriched the English language, enabling speakers to communicate complex ideas and concepts with greater precision and clarity. The development of new words and phrases that are specific to English has also facilitated the growth of the language, making it a powerful tool for universal communication and understanding.

The assimilation of varied linguistic constructs across cultural exchanges has uniquely positioned English to serve as a versatile platform for universalizing human understanding through:

- Lexical breadth - Absorbing an immense, internationally contributed vocabulary affords nuanced expression of knowledge from all disciplines and traditions.

- Grammatical flexibility - An open syntax accommodates loan translations and neologisms without loss of structural integrity, propagating foreign notions precisely.

- Conceptual interfacing - Its lexical networks associate foreign terms bidirectionally with established English meanings, building familiarity and intuitions across fields.

- Textual traditions - English scholarship has long synthesized diverse sources into standardized writings and speech accessible to broad audiences.

- Digital ubiquity - Through the Internet, English renders the world's information infrastructure navigable with minimal translation barriers.

- Lingua franca status - Its neutral role separates expressions from cultural baggage, prioritizing shared comprehension over unique linguistic origins.

By harmonizing linguistic diversity through cultural infusion, English enables ideas to transcend parochial borders. This process of progressive universalization establishes it as humanity's preeminent intellectual lever for collaborative truth-seeking on a global scale.

Transcending Cultural and Linguistic Boundaries

Discussing how English, by amalgamating various linguistic legacies, becomes a bridge for global intellectual exchange.

English has emerged as a highly effective bridge for global intellectual exchange by transcending both cultural and linguistic boundaries through its hybrid linguistic nature.

By incorporating vocabulary from the many languages and cultures it came into contact with throughout its development, English assimilated ideas, concepts and ways of thinking from diverse world regions. This enriched its semantic expressiveness while also building familiarity across linguistic divisions.

As a result, English provides shared terminology to discuss even highly specialized domains and localized traditions. Its flexibility further accommodates loan translations, preserving nuances during cross-cultural diffusion of knowledge.

English's open morphology also derives new representations of foreign notions, maintaining comprehension amid sociocultural change. And its syntactic assimilation of foreign lexicons interfaces terms bidirectionally within vast English-language networks.

Collectively, these qualities allow English to synthesize once disparate scholarly traditions into standardized texts accessible to broad global audiences. It has thus united previously partitioned intellectual discourse and come to serve as a neutral medium maximizing shared understanding over unique origins.

Through cultural infusion and linguistic syncretism, English acts as an unprecedented bridge alleviating divisions to progressively universalize human comprehension across civilizational boundaries. Its amalgamative nature cultivates more harmonized global cognition.

English, as a language, has long been recognized for its ability to transcend cultural and linguistic boundaries, fostering a platform for global intellectual exchange. This quality stems from the language's unique history, shaped by the amalgamation of various linguistic legacies.

One of the primary factors contributing to English's versatility is its Indo-European roots. The language's Germanic origins have been influenced by numerous other language families, including Latin, Greek, and French. This blend of linguistic elements has resulted in a language that is both familiar and accessible to speakers from diverse cultural backgrounds.

The borrowing of words and phrases from other languages has been a significant factor in English's development. For instance, the language has borrowed heavily from French, particularly during the Norman Conquest of England in 106- This has resulted in a significant number of French words being incorporated into the English language, especially in fields such as law, politics, and cuisine.

Similarly, English has also borrowed words from other languages such as Latin, Greek, and Arabic, which have been incorporated into the language over centuries. This has created a rich and diverse vocabulary, allowing English speakers to communicate complex ideas and concepts with precision.

The blending of linguistic legacies has not only enriched English's vocabulary but also facilitated the development of new grammatical structures. For example, the language's syntax has been influenced by both Germanic and Romance languages, resulting in a unique blend of sentence structures. This flexibility in sentence structure has enabled English speakers to express complex ideas and concepts in a variety of ways, making the language an ideal medium for intellectual exchange.

Furthermore, English's status as a global language has been fostered by its widespread use in international communication, particularly in the fields of science, technology, engineering, and mathematics (STEM). The language's neutrality, in terms of cultural and linguistic associations, has made it an attractive choice for international communication, as it is perceived as a language that is not tied to any particular culture or region.

The use of English in STEM fields has also led to the development of specialized vocabularies, which have facilitated global intellectual exchange. For instance, the language has adopted technical terms from various languages, such as "algorithm" from Arabic and "telephone" from Greek, which have become standardized in English. This shared vocabulary has enabled experts from different cultures and linguistic backgrounds to communicate and collaborate effectively, transcending cultural and linguistic boundaries.

English's ability to transcend cultural and linguistic boundaries is a result of its unique history and the amalgamation of various linguistic legacies. The language's versatility, fostered by its borrowing of words and phrases from other languages, has created a rich and diverse vocabulary, enabling English speakers to communicate complex ideas and concepts with precision. The language's neutrality and widespread use in international communication, particularly in STEM fields, have also contributed to its status as a global language, facilitating intellectual exchange across cultural and linguistic boundaries.

English's transcendent nature owes much to its historical evolution and the amalgamation of diverse linguistic legacies. This fusion has endowed the language with unparalleled versatility, allowing it to traverse cultural and linguistic barriers with ease.

The borrowing and assimilation of words and phrases from various languages have endowed English with a rich and eclectic vocabulary. This linguistic diversity empowers English speakers to articulate intricate and abstract concepts with precision, accommodating nuanced expressions that might otherwise be challenging in other languages.

Moreover, English's neutrality, reinforced by its widespread use in international communication, especially in scientific, technological, engineering, and mathematical (STEM) fields, has significantly contributed to its status as a global language. This neutrality allows English to transcend cultural biases and serve as a common ground for intellectual exchange among individuals from diverse linguistic and cultural backgrounds.

In the realm of STEM, where precise communication is imperative, English serves as a lingua franca, facilitating seamless collaboration and knowledge dissemination across borders. Its prevalence in academic journals, conferences, and research publications ensures that scientific advancements are communicated universally, enabling global participation and contribution to the progress of human knowledge.

The language's adaptability and inclusivity have further solidified its position as a tool for global intellectual exchange. English's ability to accommodate new concepts, absorb foreign terms, and evolve with changing times ensures its continued relevance and effectiveness in fostering cross-cultural communication and understanding.

Ultimately, English's capacity to transcend cultural and linguistic boundaries, rooted in its historical amalgamation and adaptability, serves as a unifying force that enables global discourse, knowledge dissemination, and collaboration across diverse communities and disciplines.

English: The Language of Science and Innovation ©

Chapter 17

Cultural Continuity and Linguistic Evolution

Cultural Continuity in Linguistic Evolution

Examining how the continuity of Latin, Greek, German, and French elements in English contributes to the evolution of thought and language.

The English language has evolved over time through the assimilation of various linguistic elements, including Latin, Greek, German, and French. This cultural continuity has played a significant role in shaping the language's vocabulary, grammar, and syntax, and has contributed to the evolution of thought and language.

- Latin: Latin has had a profound impact on the English language, particularly in the fields of law, medicine, and science. Latin words and phrases have been incorporated into English, providing a rich vocabulary for expressing complex ideas and concepts. The use of Latin roots and prefixes has also facilitated the creation of new words and terms, enabling English speakers to communicate advanced ideas and concepts.

- Greek: Greek has also had a significant influence on the English language, particularly in the fields of philosophy, literature, and the arts. Greek words and phrases have been incorporated into English, providing a vocabulary for expressing complex ideas and concepts. The use of Greek roots and prefixes has also enabled English speakers to create new words and terms, facilitating the expression of complex ideas and concepts.

- German: German has had a significant impact on the English language, particularly in the fields of technology, engineering, and business. German words and phrases have been incorporated into English, providing a vocabulary for expressing complex ideas and concepts. The use of German roots and prefixes has also facilitated the creation of new words and terms, enabling English speakers to communicate advanced ideas and concepts.

- French: French has had a significant influence on the English language, particularly in the fields of cuisine, fashion, and art. French words and phrases have been incorporated into English, providing a vocabulary for expressing complex ideas and concepts. The use of French roots and prefixes has also enabled English speakers to create new words and terms, facilitating the expression of complex ideas and concepts.

The continuity of these linguistic elements in English has contributed to the evolution of thought and language in several ways:

- Facilitating communication: The use of Latin, Greek, German, and French elements in English has enabled speakers to communicate complex ideas and concepts with greater precision and clarity.

- Enriching vocabulary: The incorporation of words and phrases from these languages has enriched the English vocabulary, providing speakers with a wider range of words and phrases to express their ideas.

- Creating new words and terms: The use of roots and prefixes from these languages has enabled English speakers to create new words and terms, facilitating the expression of complex ideas and concepts.

- Shaping grammar and syntax: The influence of these languages has also shaped the grammar and syntax of English, enabling speakers to express complex ideas and concepts in a variety of ways.

The continuity of Latin, Greek, German, and French elements in English has played a significant role in the evolution of thought and language. The incorporation of these linguistic elements has enriched the English vocabulary, facilitated communication, and enabled speakers to express complex ideas and concepts with greater precision and clarity. The use of these elements has also contributed to the creation of new words and terms, shaping the grammar and syntax of the language and facilitating the expression of complex ideas and concepts.

The continuity of linguistic elements absorbed into English has been integral to both the evolution of the language and sociocultural advancement in significant ways:

- Conceptual legacy: Enduring vocabularies from Latin, Greek ensure continuity of fundamental ideas underlying fields like science, philosophy, law that progress knowledge over generations.

- Cultural heritage: Stable Germanic and Romance language influences preserve ancestry links important for maintaining cultural identities as societies change.

- Shared cognition: Familiar linguistic constructs promote intergenerational transmission of specialized understandings, supporting creative work building on established frameworks.

- Connected epistemology: Etymological ties between English words and those in other languages encourage tracing ideas across disciplines and through history.

- Linguistic ties: Retaining cognates anchors English within wider Indo-European linguistic relations, maintaining sociohistorical contextualization.

- Flexible adaptation: Yet continual borrowing enables English to dynamically adapt itself by synthesizing foreign constructs congruent with evolving thought patterns.

The strategic integration of absorbed linguistic elements within English not only bolsters its linguistic stability but also nurtures a continuum of cultural heritage. This strategic assimilation plays a pivotal role in preserving past wisdom while fostering a platform for future progress, rooted in a shared knowledge foundation.

The process of absorbing linguistic elements from various sources strategically embeds these components within the framework of English. This integration is not merely a superficial inclusion but a deliberate amalgamation that lends stability to the language. By incorporating these elements strategically, English maintains a cohesive structure while embracing the richness of diverse cultural and linguistic influences.

This linguistic stability, grounded in the assimilation of varied elements, facilitates cultural continuity. It serves as a bridge between the past and the future, preserving the accumulated wisdom, insights, and cultural nuances encapsulated within these absorbed linguistic elements. This continuity ensures that historical knowledge and cultural legacies persist within the fabric of the language, providing a foundation for future generations to build upon.

Furthermore, this shared knowledge foundation nurtures a collective understanding and appreciation of diverse cultural perspectives. By anchoring past wisdom within the language, English becomes a repository of collective human experiences, fostering a sense of interconnectedness and shared heritage among its speakers.

This continuum of cultural continuity not only preserves historical wisdom but also paves the way for future progress. By providing a stable platform built upon accumulated knowledge, English becomes a fertile ground for innovation, facilitating the exchange and evolution of ideas across generations and cultural boundaries.

The strategic permanence of absorbed linguistic elements in English serves as a conduit for cultural continuity, ensuring the preservation of past wisdom and nurturing an environment conducive to future progress. It fosters a shared knowledge foundation that transcends time, enriching the language and its speakers with a sense of heritage and a trajectory for continuous growth and innovation.

Influence on Thought Patterns and Ideation

Discussing how the amalgamation of linguistic roots shapes thought patterns and ideation processes within English-speaking societies.

The amalgamation of linguistic roots in English has had a significant impact on thought patterns and ideation processes within English-speaking societies. The influence of various languages, such as Latin, Greek, German, and French, has contributed to the development of a rich and diverse vocabulary, which in turn has shaped the way people think and communicate.

- Latin: Latin has had a profound influence on the English language, particularly in the fields of law, medicine, and science. The incorporation of Latin roots and prefixes has enabled English speakers to communicate complex ideas and concepts with precision and clarity. Latin has also contributed to the development of a systematic and logical approach to thinking, which is reflected in the structured nature of English grammar and sentence structure.

- Greek: Greek has had a significant impact on English vocabulary, particularly in the fields of philosophy, literature, and the arts. Greek roots and prefixes have enabled English speakers to express complex ideas and concepts related to these fields with greater nuance and depth. The influence of Greek has also contributed to the development of a more analytical and abstract approach to thinking, which is reflected in the use of Greek-derived words in English to express abstract concepts.

- German: German has had a significant impact on English vocabulary, particularly in the fields of technology, engineering, and business. German roots and prefixes have enabled English speakers to communicate complex ideas and concepts related to these fields with greater precision and clarity. The influence of German has also contributed to the development of a more practical and hands-on approach to thinking, which is reflected in the use of German-derived words in English to express technical and scientific concepts.

-   French: French has had a significant impact on English vocabulary, particularly in the fields of cuisine, fashion, and art. French roots and prefixes have enabled English speakers to express complex ideas and concepts related to these fields with greater nuance and sophistication. The influence of French has also contributed to the development of a more aesthetic and creative approach to thinking, which is reflected in the use of French-derived words in English to express artistic and cultural concepts.

The amalgamation of linguistic roots in English has also influenced thought patterns and ideation processes in several ways:

-   Facilitating communication: The use of a diverse range of linguistic roots and prefixes has enabled English speakers to communicate complex ideas and concepts with greater precision and clarity.

- Enriching vocabulary: The incorporation of words and phrases from various languages has enriched the English vocabulary, providing speakers with a wider range of words and phrases to express their ideas.

- Encouraging creativity: The influence of languages such as French has contributed to the development of a more aesthetic and creative approach to thinking, enabling English speakers to express complex ideas and concepts in a more nuanced and sophisticated way.

- Shaping grammar and syntax: The influence of various languages has also shaped the grammar and syntax of English, enabling speakers to express complex ideas and concepts in a variety of ways.

The amalgamation of linguistic roots in English has had a significant impact on thought patterns and ideation processes within English-speaking societies. The incorporation of Latin, Greek, German, and French roots and prefixes has enabled English speakers to communicate complex ideas and concepts with greater precision and clarity, and has contributed to the development of a rich and diverse vocabulary. The influence of these languages has also shaped the grammar and syntax of English, enabling speakers to express complex ideas and concepts in a variety of ways, and has encouraged creativity and nuance in expression.

The fusion of diverse linguistic roots has significantly shaped thought patterns and ideation processes among English speakers in several ways:

- Vocabulary influences conceptualization. Absorbing lexicons gave access to new ideas and frameworks that permeated mindsets (e.g. Greek philosophical terms).

- Etymology impacts associations. Shared roots between words activate related concepts, influencing how ideas are networked and inferences made.

- Grammar facilitates certain cognitions. Analytic structures like SVO favor categorization while inflected languages promote relational thinking.

- Discourse conventions emerge. Integrating scholarly traditions birthed new text genres and argument forms that molded rational processes.

- Figurative devices become tools for abstract thought. Semantic shifts and metaphorical terms embedded new ways to model complex topics.

- Field-specific lexicons guided expertise. Adopting nuanced lexicons nudged focus toward compatible domains and methodologies.

- Neologisms capture emerging notions. Coining terms in step with change assisted comprehending technological/social disruptions.

- Register differences modify perspectives. Juggling scales of formality enhanced adaptability to varied contexts and audiences.

English's hybrid linguistic inheritance significantly tuned its speakers' cognitive proclivities and enabled innovative modes of creatively advancing knowledge. This stands as a catalyst that intricately shapes the cognitive tendencies of its speakers while fostering innovative pathways for advancing knowledge.

The amalgamation of diverse linguistic elements within English has indeed molded the cognitive landscape of its speakers. This hybrid inheritance has honed cognitive proclivities, influencing how individuals perceive, process, and articulate thoughts. The language's diverse roots offer a multifaceted lens through which concepts are understood, encouraging a holistic approach to comprehension and problem-solving.

Moreover, this linguistic hybridity has stimulated innovative modes of knowledge advancement. English, with its blended inheritance, prompts creative thinking and ideation. The fusion of linguistic legacies within the language encourages novel approaches to expressing ideas, fostering a mindset that values experimentation and unconventional connections between concepts.

The inherent flexibility of English, born from its hybrid linguistic makeup, allows for the creation of new pathways in knowledge exploration. Its adaptability to absorb and assimilate concepts from diverse domains facilitates interdisciplinary exploration and the synthesis of ideas from different fields. This fosters an environment where cross-disciplinary innovations thrive, propelling the boundaries of knowledge forward.

Furthermore, the cognitive tuning instilled by English's hybrid inheritance encourages a penchant for nuanced expression and abstract thought. This proficiency in navigating diverse linguistic influences equips individuals with the ability to articulate complex ideas with depth and precision, paving the way for innovative contributions across various domains.

English's hybrid linguistic inheritance not only shapes the cognitive tendencies of its speakers but also fosters an environment conducive to innovative knowledge advancement. Its amalgamation of linguistic elements acts as a catalyst, nurturing creative thinking, interdisciplinary exploration, and nuanced expression, ultimately contributing to the ongoing evolution and enrichment of human knowledge.

Chapter 18

The Triumvirate of Linguistic Influence

The synergy of Latin, Greek, and Germanic roots within the English language is indeed a remarkable example of linguistic evolution and intellectual enrichment. The English language has been shaped by various influences, including the Latin language, which was introduced to England by the Normans after the Norman Conquest of 106- Latin, being the language of the Church and the language of learning, had a profound impact on the English language, particularly in the fields of law, medicine, and science.

The Greek language also played a significant role in the development of English, particularly in the fields of philosophy, literature, and the arts. Greek words and phrases were incorporated into English, providing a vocabulary for expressing complex ideas and concepts. The Germanic roots of English, which include Old English, Middle English, and Modern English, have also had a significant impact on the language, particularly in the fields of technology, engineering, and business.

The synergy of these linguistic roots has resulted in a rich and diverse vocabulary, which has enabled English speakers to communicate complex ideas and concepts with precision and clarity. The blending of Latin, Greek, and Germanic roots has also contributed to the development of a systematic and logical approach to thinking, which is reflected in the structured nature of English grammar and sentence structure.

Furthermore, the influence of these languages has also shaped the grammar and syntax of English, enabling speakers to express complex ideas and concepts in a variety of ways. The use of Latin, Greek, and Germanic roots has also encouraged creativity and nuance in expression, allowing English speakers to convey subtle shades of meaning and express complex thoughts and emotions.

The synergy of Latin, Greek, and Germanic roots within the English language is a testament to the power of linguistic evolution and intellectual enrichment. The blending of these languages has resulted in a rich and diverse vocabulary, a systematic approach to thinking, and a flexible and expressive language that continues to evolve and adapt to the needs of its speakers.

The synergy created by the fusion of Latin, Greek, and Germanic roots is truly remarkable. A few additional thoughts:

- The combination of inflected Latin/Greek vocabulary provided precision and nuance, while Germanic grammar lent flexibility and clarity. This created an optimized vehicle for complex ideas.

- Core concepts from philosophy, science, governance were grafted into English via Latin/Greek, planting seminal seeds that flourished through open hybridization.

- Germanic influence retained cultural identity while Latin/Greek injection propelled participation in continental Renaissance. Together they balanced continuity and progress.

- Fluid absorption of new notions mirrored humankind's intellectual evolution, ensuring relevance. English naturally adapts to emerging fields.

- Shared Indo-European ancestry enhanced legibility of imported lexicons, lowering cognitive load versus purely foreign frameworks.

- Today, this fortuitous historical synergy leaves English overwhelmingly well-equipped to facilitate global exchange of perspectives.

The artful linguistic synthesis truly constitutes one of history's great intellectual multiplications. It establishes English as humanity's foremost common medium of learning and understanding.

Legacy of Latin Precision

Latin, renowned for its precision and clarity, injected English with a reservoir of scientific, legal, and scholarly terminologies. The meticulous structure of Latin allowed for the preservation of intricate concepts and scholarly discourse, enhancing the precision and technicality of English in specialized fields.

The influence of Latin on the English language has been profound, particularly in the fields of science, law, and scholarship. Latin's precision and clarity have contributed to the development of a rich and specialized vocabulary in English, allowing for the precise expression of complex concepts and ideas.

In science, Latin has provided English with a wealth of technical terminologies that have enabled scientists to communicate their findings with precision and accuracy. For example, Latin words such as "species," "genus," and "phylum" are used in biology to describe the classification of living organisms, while words such as "atom," "molecule," and "chemical reaction" are used in chemistry to describe the building blocks of matter and the processes that govern their interactions.

In law, Latin has contributed to the development of legal terminology that is used to describe complex legal concepts and principles. Latin words such as "habeas corpus," "pro bono," and "mens rea" are used in legal contexts to describe concepts such as the right to a fair trial, the duty to act in the best interests of a client, and the mental state of a criminal defendant.

In scholarship, Latin has provided English with a range of words and phrases that are used to describe academic disciplines and concepts. For example, Latin words such as "philosophia," "historia," and "literatura" are used to describe the disciplines of philosophy, history, and literature, respectively.

The meticulous structure of Latin has also influenced the way in which English words are formed and used. Latin's system of declension, which governs the way in which nouns and pronouns are inflected to indicate grammatical case, has influenced the development of English grammar and syntax. For example, English words such as "dog" and "cat" are inflected to indicate grammatical case, with forms such as "dogs" and "cats" used to indicate plural nouns.

In addition, Latin's system of conjugation, which governs the way in which verbs are inflected to indicate tense, mood, and person, has influenced the development of English verb conjugation. English verbs such as "to go," "to eat," and "to run" are conjugated to indicate tense, mood, and person, with forms such as "went," "ate," and "ran" used to indicate past tense.

The legacy of Latin precision in English has resulted in a language that is capable of expressing complex ideas and concepts with clarity and accuracy. The specialized vocabulary and grammatical structures that Latin has contributed to English have enabled scholars, scientists, and legal professionals to communicate their ideas with precision and technicality, making English a powerful tool for intellectual discourse.

We've highlighted Latin's tremendous legacy in enriching English's potential for precise expression. Some additional thoughts on how Latin shaped English's technical capacity:

- By absorbing Latin's vast stock of descriptive terms, English gained access to the conceptual frameworks and taxonomies underlying advanced Roman knowledge in law, governance, philosophy, medicine and more.

- From Latin prefixes, suffixes and grammar rules, English acquired the morphological tools to precisely define new notions or subtly alter established meanings as fields evolved.

- Scholarship produced in Latin provided templates for specialized syntax, objectivity and argumentative styles that carried over as linguistic conventions in English writings.

- Cottage industries in translating Latin texts into English augmented localized expertise by disseminating discoveries and methodologies around Europe.

- Neo-Latin efflorescence during the Renaissance infused English with waves of innovative terminology cementing new scholarly paradigms and techniques.

- Even now, over half of English's core vocabulary stems from Latin, endowing it with an unrivaled precision for intellectual discourse across STEM, social sciences and humanities.

Latin's legacy indeed casts a profound influence on the precision and depth of expression within English, rendering it unparalleled in systematically conveying nuanced and complex ideas.

The meticulous nature of Latin, renowned for its precision and clarity, has left an indelible mark on English. The assimilation of Latin roots, phrases, and structures has bestowed upon English a reservoir of specialized terminology and syntactic structures, facilitating the meticulous expression of intricate concepts.

English, through its incorporation of Latin elements, inherits a precision that enables speakers to articulate complex ideas with meticulous detail and accuracy. Latin-derived vocabulary not only enhances the language's lexicon but also provides succinct and precise terminology for expressing abstract or technical concepts in various fields, including law, medicine, science, and philosophy.

Furthermore, Latin's legacy in English syntax and sentence structures contributes to the systematic conveyance of complex ideas. The adoption of Latin-inspired syntactic patterns allows for the construction of intricate sentences that maintain logical coherence and clarity, enabling the conveyance of multifaceted thoughts with precision.

This legacy of meticulous expression from Latin empowers English speakers to navigate intricate concepts with a level of detail and accuracy that few languages can match. The linguistic legacy inherited from Latin enriches English with a depth of vocabulary and syntactic structures, enabling the systematic conveyance of nuanced and complex ideas across various domains of knowledge and discourse.

Greek Philosophical Abundance

From philosophy to science, Greek contributions have bestowed English with an array of philosophical, scientific, and mathematical terminologies. The philosophical depth of Greek thought found its way into English, enriching its lexicon with words that encapsulate nuanced ideas and abstract concepts.

The Greek philosophical tradition has had a profound impact on the English language, contributing a wealth of terminology that has enriched its lexicon and enabled the expression of complex ideas and abstract concepts. From the works of Plato and Aristotle to the teachings of the Stoics and Epicureans, Greek philosophy has shaped the way we think and communicate.

In philosophy, Greek concepts such as "ethos," "pathos," and "logos" have been incorporated into English, providing a vocabulary for discussing the art of persuasion and the nature of argumentation. Other philosophical terms, such as "metaphysics," "epistemology," and "ontology," have also been borrowed from Greek, allowing English speakers to engage in sophisticated philosophical inquiry.

In science, Greek words such as "biology," "physics," and "chemistry" have been adopted into English, providing a foundation for the scientific study of the natural world. Greek mathematical terms, such as "geometry," "algebra," and "trigonometry," have also been incorporated into English, enabling mathematicians and scientists to describe complex mathematical concepts with precision.

The influence of Greek on the English language can also be seen in the many Latinized Greek words that have been incorporated into English. Words such as "philosophy," "psychology," and "sociology" are all derived from Greek roots and have been adopted into English, providing a vocabulary for discussing various fields of study.

In addition to its direct contributions to the English language, Greek has also had an indirect influence on English through its influence on Latin. Many Latin words that were borrowed into English have Greek roots, and the meanings of these words have been shaped by their original Greek context. For example, the Latin word "physica," which means "natural science," is derived from the Greek word "phusikē," which means "natural philosophy."

The impact of Greek on the English language has been profound, providing a rich and diverse vocabulary that has enabled English speakers to engage in sophisticated intellectual discourse. The philosophical depth and scientific rigor of Greek thought have found their way into English, making it a powerful tool for communicating complex ideas and abstract concepts.

Further aspects of Greek's influence:

- Core philosophical, rhetorical and analytical terms introduced by Socrates, Plato and Aristotle formed the basis of nuanced argumentation in English.

- Fields like mathematics, astronomy and natural sciences bore the earliest seeds of empiricism and theoretical modeling via Greek imports.

- Subtleties of emotion, aesthetics, spirituality and metaphysical discourse gained expression through exposure to Greek literature and mythology.

- Revivals of interest in Classical works, like during the English Renaissance, stimulated new lexical borrowing and innovative conceptual blending.

- Neoplatonic, gnostic and hermetic strands of later Greco-Roman thought found voice in English, enlarging its semantic ambit.

- Greek impact expanded through chains of Classical learning, as ideas percolated into English indirectly via Latin, Arabic and other intermediaries.

Indeed, few languages can rival the complexity and depth Greek endowed English with to tackle nuanced epistemological problems. Its philosophical abundance proved enormously catalytic.

The influence of Greek on English has been pivotal in endowing the language with the capacity to address nuanced epistemological complexities. Greek's profound philosophical abundance has served as a catalytic force in shaping English's ability to navigate intricate philosophical inquiries.

The rich philosophical legacy inherited from Greek has significantly enriched English vocabulary, infusing it with a plethora of terms that encapsulate profound philosophical concepts. Words derived from Greek, such as "ontology," "epistemology," and "metaphysics," exemplify this inheritance, enabling English speakers to articulate intricate philosophical ideas with precision and depth.

Moreover, Greek philosophy's profound impact extends beyond mere vocabulary enrichment. It has fundamentally shaped the conceptual framework and thought processes within English-speaking cultures. The depth of Greek philosophical thought serves as a guiding force, influencing how English speakers approach and dissect epistemological problems.

The catalytic effect of Greek philosophy on English lies in its stimulation of intellectual curiosity and critical thinking. The legacy of Greek philosophical concepts, theories, and methodologies encourages a rigorous approach to analyzing and addressing nuanced epistemological dilemmas, fostering a culture of inquiry and exploration within English-speaking societies.

Furthermore, the assimilation of Greek philosophical concepts into English has facilitated interdisciplinary connections, allowing for the integration of philosophical ideas into various domains. This interdisciplinary exchange of ideas broadens the scope of inquiry, fostering innovative approaches to addressing complex epistemological problems.

Greek's profound philosophical abundance has been instrumental in equipping English with the vocabulary, conceptual framework, and intellectual rigor necessary to grapple with nuanced epistemological challenges. Its enduring legacy continues to catalyze a culture of critical thinking and inquiry, enabling English to navigate and unravel intricate philosophical inquiries with depth and precision.

Germanic Foundation and Expressiveness

The Germanic roots of English, with their innate expressiveness and emotive power, form the foundation of its linguistic structure. These roots infuse English with a visceral quality, enabling it to convey emotions, narrative depth, and everyday communication with a richness that resonates deeply within its speakers.

The Germanic roots of English have played a significant role in shaping the language's linguistic structure and expressive capabilities. The Germanic languages, including Old English, Old Norse, and Gothic, have contributed a wealth of vocabulary and grammatical elements that have helped to create a language that is both functional and emotive.

One of the key characteristics of Germanic languages is their innate expressiveness and emotive power. These languages have a rich collection of words and phrases that convey emotions, moods, and atmospheres, which has been integrated into English. This expressiveness is evident in the many Germanic-derived words and phrases that are used in English to describe emotions, such as "angst," "fear," "joy," and "sorrow."

Moreover, the Germanic roots of English have also influenced the language's grammatical structure, particularly in the area of sentence formation. Germanic languages are known for their flexible word order, which allows for a range of expressive possibilities. This flexibility has been incorporated into English, enabling speakers to convey nuanced shades of meaning through the arrangement of words in a sentence.

Another way in which the Germanic roots of English have contributed to its expressiveness is through the use of inflectional endings. Germanic languages are characterized by a system of inflectional endings that indicate grammatical case, tense, and number. English has inherited this system, which allows speakers to convey subtle shades of meaning through the use of different inflectional endings.

Furthermore, the Germanic roots of English have also influenced the language's vocabulary, particularly in the areas of everyday communication and narrative storytelling. Germanic languages have a rich collection of words that describe everyday objects, actions, and experiences, which have been incorporated into English. This has enabled English speakers to communicate complex ideas and narratives with a level of detail and nuance that is unique among languages.

The Germanic roots of English have played a crucial role in shaping the language's linguistic structure and expressive capabilities. The innate expressiveness and emotive power of Germanic languages have been integrated into English, enabling speakers to convey emotions, narrative depth, and everyday communication with a richness that resonates deeply within its speakers. The flexible word order, inflectional endings, and vocabulary of Germanic languages have all contributed to the creation of a language that is both functional and expressive, making English a powerful tool for communication and storytelling.

Germanic roots have played a significant role in making English such a vibrant, resonant language. A few additional thoughts:

- The core Anglo-Saxon vocabulary gives English a down-to-earth quality and directness ideal for storytelling, poetry and oral tradition.

- Germanic grammar, with its flexible word order, allowed the language to evolve in an open and dynamic way to absorb new influences.

- Each Germanic language brought its own expressive nuances - Norse imagery, unconquered spelling/pronunciation variations.

- Everyday terms for the home, community and natural world retained ethnic connection to landscape and lived experience.

- Lexical items like "fore-" prefixes enhanced description of processes, events unfolding through time.

- Resonant phonaesthetics like onomatopoeia give English mnemonic and emotional features.

- Cultural elements from England, Germany, Scandinavia blended into a rich expressive substrate.

Indeed, the Germanic linguistic DNA has left an indelible mark on English, endowing it with qualities of earthiness, vivacity, and emotional immediacy that resonate profoundly among global English speakers even today.

The infusion of Germanic roots within English imparts a distinct character marked by earthy authenticity and directness. Words derived from Germanic origins often convey a sense of groundedness, reflecting the raw, tangible aspects of life. This earthiness lends a sense of robustness and authenticity to English expressions, allowing for direct and straightforward communication that connects deeply with speakers across cultures.

Moreover, the Germanic influence contributes to the vivacity and dynamism inherent in English. Words rooted in Germanic origins often carry a sense of energy and vigor, allowing for vivid and impactful expressions. This vivacity adds a layer of immediacy to the language, enabling English speakers to convey emotions, experiences, and actions with a sense of liveliness and intensity.

Additionally, the emotional immediacy embedded in Germanic-influenced terms allows English to convey emotions and sentiments with a raw and immediate quality. These words often encapsulate deep emotional nuances, enabling speakers to express feelings with a degree of intensity and vividness that resonates profoundly with listeners and readers.

The enduring influence of the Germanic linguistic DNA on English persists as a vital element in the language's expressive palette. It contributes to the richness and diversity of English, enabling speakers to communicate with a blend of earthiness, vivacity, and emotional immediacy that transcends cultural boundaries, forging connections and resonating deeply with audiences worldwide.

The Harmonious Fusion

The amalgamation of these linguistic heritages within English creates a mosaic of expression, where precision meets philosophical depth and emotive power. This fusion not only expands the language's breadth but also nurtures a fertile ground for intellectual exploration and innovation.

The blending of Latin, Greek, and Germanic linguistic traditions within English has resulted in a language that boasts a unique fusion of precision, philosophical depth, and emotive power. This mosaic of expression allows English speakers to convey complex ideas and emotions with nuance and sophistication, making it an ideal language for intellectual exploration and innovation.

The Latin influence, with its focus on precision and clarity, provides a foundation for technical and scientific communication. The language's Germanic roots, with their emphasis on emotive power and storytelling, add depth and nuance to English, enabling speakers to convey complex emotions and narratives with ease. The philosophical depth of Greek, with its focus on abstract concepts and ideas, further enhances English's ability to convey complex thought and expression.

This fusion of linguistic traditions not only expands the language's breadth but also creates a fertile ground for intellectual exploration and innovation. English speakers can draw upon a vast array of words, phrases, and grammatical structures to express their ideas, making it a versatile language that can adapt to a wide range of contexts and disciplines.

Moreover, the blending of these linguistic traditions has also contributed to the development of a rich and diverse cultural heritage. English literature, for example, draws heavily from the language's Germanic and Latin roots, with works such as Beowulf and The Canterbury Tales showcasing the language's ability to convey complex narratives and emotions. The language's philosophical depth has also made it a popular choice for philosophers and thinkers, with works such as Shakespeare's Hamlet and Milton's Paradise Lost exploring profound themes and ideas.

The harmonious fusion of Latin, Greek, and Germanic linguistic traditions within English has created a language that is both precise and emotive, with a rich cultural heritage that continues to inspire intellectual exploration and innovation.

The harmonious fusion of Greek, Latin and Germanic elements has truly forged English into an unparalleled linguistic instrument. A few more thoughts:

- Each contributor strengthened different dimensions, and together they optimize English for varied modes of cogent, impactful communication.

- By preserving ancestral connections while also globalizing thoughtscapes, English synchronizes continuity and progress.

- Its flexible constitution seamlessly integrates outside constructs, broadcasting innovations universally.

- English's hybrid vigor emanates from a negotiated balance of explicitness and nuance, analysis and emotive aesthetic.

- No language matches its spectrum, reconciling logos, ethos and pathos for broadest possible discourse.

- And this dynamic synthesis shows no signs of closure - English will forever evolve by synthesis anew.

English, with its rich linguistic heritage, serves as a catalyst for limitless intellectual prospects. This quality of English, welcoming diverse perspectives and fostering constructive intersections, indeed stands as a linguistic triumph.

The amalgamation of linguistic legacies within English creates a fertile ground where ideas from diverse cultural, historical, and disciplinary backgrounds intersect and flourish. This diversity in linguistic influences provides a vast reservoir of perspectives and concepts, encouraging a culture of intellectual exploration and innovation.

English's ability to embrace varied linguistic elements enables it to serve as a melting pot for ideas, allowing for the constructive intersection of diverse viewpoints. This intersection cultivates an environment where different disciplines converge, facilitating interdisciplinary dialogues and paving the way for novel insights and discoveries.

Moreover, English's inclusive nature, rooted in its amalgamated linguistic past, fosters an open and welcoming space for diverse perspectives to coexist and collaborate. This inclusivity encourages dialogue and exchange, enabling the synthesis of ideas and the emergence of innovative solutions to complex problems.

This linguistic triumph of English lies in its capacity to act as a bridge, connecting different cultures, disciplines, and ideas. It provides a platform where intellectual exploration knows no bounds, nurturing an environment conducive to innovation and paving the way for continuous advancement and growth in knowledge.

The phrase eloquently captures how English, through its diverse linguistic heritage, serves as fertile soil for intellectual exploration and innovation. Its ability to welcome diverse perspectives and facilitate constructive intersections stands as a testament to its pivotal role in fostering a thriving global discourse and shaping the trajectory of intellectual progress.

Impact on Cognitive Synthesis

This triumvirate of linguistic influence in English enhances cognitive synthesis, offering a diverse spectrum of linguistic tools to articulate thoughts, theories, and discoveries. It facilitates a seamless integration of ideas from various disciplines, propelling interdisciplinary studies and fostering a holistic approach to knowledge.

The intersection of Latin, Greek, and Germanic linguistic traditions in English has a profound impact on cognitive synthesis, providing a diverse array of linguistic tools that enable speakers to articulate complex thoughts, theories, and discoveries. This multifaceted linguistic heritage allows for a seamless integration of ideas from various disciplines, fostering interdisciplinary studies and a holistic approach to knowledge.

The Latin influence, with its focus on precision and clarity, provides a foundation for logical reasoning and analytical thinking. The language's Germanic roots, with their emphasis on emotive power and storytelling, add depth and nuance to English, enabling speakers to convey complex emotions and narratives with ease. The philosophical depth of Greek, with its focus on abstract concepts and ideas, further enhances English's ability to convey complex thought and expression.

This synthesis of linguistic traditions enables English speakers to draw upon a vast array of words, phrases, and grammatical structures to express their ideas, making it a versatile language that can adapt to a wide range of contexts and disciplines. It allows for the creation of complex sentences that can convey multiple ideas and relationships, facilitating the expression of sophisticated thoughts and arguments.

Moreover, the blending of these linguistic traditions has also contributed to the development of a rich and diverse cultural heritage. English literature, for example, draws heavily from the language's Germanic and Latin roots, with works such as Beowulf and The Canterbury Tales showcasing the language's ability to convey complex narratives and emotions. The language's philosophical depth has also made it a popular choice for philosophers and thinkers, with works such as Shakespeare's Hamlet and Milton's Paradise Lost exploring profound themes and ideas.

The intersection of Latin, Greek, and Germanic linguistic traditions in English has a profound impact on cognitive synthesis, offering a diverse spectrum of linguistic tools that enable speakers to articulate complex thoughts, theories, and discoveries. It facilitates a seamless integration of ideas from various disciplines, propelling interdisciplinary studies and fostering a holistic approach to knowledge.

English promotes cognitive advancements through its symbiotic linguistic roots. A few more insights:

- Its expansive vocabulary flexibly depicts specialized and abstract concepts, bridging epistemic divides.

- Shared etymological structures encourage analogical reasoning across fields.

- Grammatical openness lets hybridized constructs clarify multi-faceted relationships.

- Literary conventions translated diverse thought traditions into accessible yet rigorous syntax.

- Cultural infusion continually widens English's contextual grasp, revealing unity within diversity.

- Neologizing captures emergent, intersectional ideas right at their inception points.

- Users intuitively blend linguistic registers for optimal interdisciplinary expression.

English exactly mirrors humankind's growing tendency to synthesize information holistically rather than compartmentalize. This positions it perfectly as our foremost shared tool for integrating universal knowledge. Certainly it has profoundly catalyzed more integrated cognitive processes!

Indeed, English serves as a reflection of humankind's evolving inclination toward holistic information synthesis rather than compartmentalization. This quality positions English as the foremost shared tool for integrating universal knowledge, profoundly catalyzing more integrated cognitive processes among its speakers.

The evolution of English mirrors a global shift towards a holistic approach in processing and synthesizing information. The language's amalgamation of diverse linguistic elements allows for the integration of multifaceted concepts and perspectives, aligning with humanity's inclination towards holistic understanding rather than isolating information into compartments.

English's flexibility and adaptability facilitate the seamless integration of universal knowledge across diverse domains and disciplines. It acts as a conduit for interdisciplinary connections, enabling individuals to synthesize information from different sources and fields, fostering a more comprehensive understanding of complex phenomena.

Moreover, English's role as a shared tool for integrating universal knowledge stems from its ability to accommodate diverse perspectives and insights. The language's inclusive nature encourages a holistic view that embraces various cultural, scientific, and philosophical viewpoints, promoting a more integrated and interconnected approach to cognition.

The language's capacity to bridge gaps between different forms of knowledge facilitates a more integrated cognitive process among its users. English encourages individuals to draw connections, identify patterns, and synthesize information holistically, nurturing a cognitive landscape that values interconnectedness and synthesis over compartmentalization.

English's alignment with humankind's inclination towards holistic information synthesis positions it as a primary tool for integrating universal knowledge. Its ability to foster more integrated cognitive processes among its speakers underscores its role in shaping a collective mindset that embraces interconnectedness and holistic understanding, contributing significantly to the advancement and integration of global knowledge.

Cultural Continuity and Adaptability

The cultural continuity and adaptability of Latin, Greek, and Germanic elements within English are a testament to the dynamic nature of language. Language is not a static entity, but rather a constantly evolving medium that adapts to the needs of its speakers. The presence of these elements in English demonstrates how language can absorb and assimilate various cultural influences, creating a rich and diverse linguistic landscape.

The Latin influence, for instance, has contributed to the development of English as a language of science, technology, and academia. The language's Germanic roots have shaped its grammar, vocabulary, and idiomatic expressions, making it a versatile tool for storytelling and communication. The philosophical depth of Greek has enriched English's ability to convey complex ideas and abstract concepts.

Moreover, the adaptation of Latin, Greek, and Germanic elements in English has allowed the language to remain relevant and effective in various contexts. English has become a lingua franca, a common language that bridges cultural and geographical divides. Its adaptability has enabled it to become a powerful tool for communication, commerce, and creative expression.

The continuity of these elements also highlights the importance of understanding the historical and cultural contexts in which language evolves. By studying the development of English, we can gain insights into the social, political, and cultural shifts that have shaped our world. We can appreciate how language reflects and influences human culture, and how it continues to adapt to meet our changing needs.

The cultural continuity and adaptability of Latin, Greek, and Germanic elements within English demonstrate the versatility and resilience of language. They show how language can absorb and assimilate diverse cultural influences, creating a rich and dynamic linguistic landscape that resonates across time and disciplines. The evolution of English serves as a reminder of the importance of understanding the historical and cultural contexts that shape language, and how language continues to adapt to meet our changing needs.

The balance English maintains between continuity and adaptability through cultural infusion indeed mirrors its biological nature and ability to both preserve ancestral traits and evolve in response to changing environmental pressures. A few more thoughts:

- Retaining familiar constructs, genres and scholarship traditions provides cognitive anchors as societies progress.

- Yet flexible word formation readily assimilates globalizing terminology, maintaining linguistic relevance.

- Recurrent borrowing cycles stabilize long-term understanding while integrating disruptive ideas.

- Multilayered histories rooted diversity but epistemic ties to other languages facilitate connections.

- Open constitution harmonizes descent and innovation, cultural memory with modernity.

- Etymological depth imbues ongoing renewal with deeper contextual purport.

- Stewardship ensures English remains optimally fit for purpose across millennia.

English language and cultures co-evolve in adaptive reciprocity. This dynamic congruence secures English's longevity as the natural cross-temporal conduit for ever-advancing human knowledge. Truly it is evolution's gift to global cognition.

The symbiotic relationship between the English language and diverse cultures epitomizes an adaptive reciprocity, fostering a dynamic congruence that secures English's longevity as the natural conduit for advancing human knowledge across time. This interplay stands as evolution's gift to global cognition.

The co-evolution of English and cultures illustrates a reciprocal relationship where the language adapts and evolves in tandem with the diverse cultural landscapes it encounters. This dynamic congruence ensures the language remains responsive to societal shifts, embracing new ideas, expressions, and perspectives as it integrates them into its fabric.

This adaptive reciprocity between language and cultures safeguards English's longevity as a cross-temporal conduit for human knowledge. English, through its evolving nature, becomes a repository of accumulated wisdom, blending past insights with contemporary innovations and paving the way for future advancements.

Moreover, this symbiotic relationship ensures that English remains relevant and adaptable in an ever-changing world. As cultures evolve, so does the language, reflecting the multifaceted experiences and diverse perspectives of its speakers. This adaptability secures English's position as a dynamic tool for communication, intellectual exchange, and knowledge dissemination across temporal boundaries.

The synergy between English and cultures serves as a cornerstone for global cognition, fostering a shared platform where diverse ideas intersect and evolve. This dynamic interplay amplifies the language's capacity to serve as a vessel for the continuous advancement of human knowledge, transcending temporal limitations and nurturing a collective reservoir of wisdom and innovation.

The reciprocal relationship between the English language and cultures is a testament to evolution's gift to global cognition. This symbiotic evolution ensures English's endurance as a conduit for ever-advancing human knowledge, embodying a dynamic congruence that propels the collective evolution of global intellect and understanding.

The convergence of Latin, Greek, and Germanic roots in English has had a profound impact on the language's vocabulary, but it has also imbued it with a multidimensional eloquence that facilitates the exchange and evolution of ideas across various fields of human inquiry. This multidimensionality allows English to be used as a conduit for the expression and dissemination of complex ideas, fostering a deeper understanding and dialogue between different disciplines and cultures.

The Latin influence, with its focus on precision and clarity, has contributed to the development of technical and scientific terminology, enabling English to accurately convey complex concepts and theories. The Greek influence, with its emphasis on philosophical and abstract ideas, has enriched English's ability to express profound and nuanced thoughts, while the Germanic influence, with its focus on storytelling and narrative, has given English a unique ability to convey emotions and tell compelling stories.

The blending of these linguistic traditions has created a language that is both precise and expressive, allowing it to adapt to a wide range of contexts and disciplines. English has become a lingua franca, a common language that bridges cultural and geographical divides, facilitating the exchange of ideas and knowledge across borders.

Moreover, the multidimensional eloquence of English has enabled it to serve as a conduit for the evolution of ideas, as it can express complex concepts and theories in a way that is both accessible and profound. This has allowed English to become a powerful tool for communication, commerce, and creative expression, and has played a significant role in shaping the modern world.

The convergence of Latin, Greek, and Germanic roots in English has not only augmented its vocabulary but has also imbued it with a multidimensional eloquence that serves as a conduit for the exchange and evolution of ideas across diverse realms of human inquiry. This unique blend of linguistic traditions has made English a versatile and powerful language, capable of adapting to a wide range of contexts and disciplines, and has played a significant role in shaping the modern world.

The coming together of these diverse linguistic roots has indeed imbued English with true multidimensional eloquence that perfectly suits its role as a premier conduit for the exchange and evolution of ideas. A few additional insights:

- The nuanced expressiveness across technical, literary, emotive registers optimizes communication in any context or field.

- Its capacious lexicon interfacing multifaceted networks of meaning enhances coherent conceptualization.

- Fluid syntax adapts narration of knowledge seamlessly between traditional and interdisciplinary realms.

- Shared etymological structures aid transference of frameworks across languages and epochs.

- Cultural history fosters appreciation of diversity within common pursuits of truth.

- Proven capacity to synthesize foreign constructs expands the horizons of what can be shared.

- Continual renewal maintains comprehension amid changing paradigms.

English indeed transcends limitations, offering itself as a language that serves as a gateway for all seekers of understanding. Its rich past serves as a prologue to an endlessly fertile future of cognitive progress, fostering collaboration without borders. This remarkable success story of English embodies the harmonious amalgamation of both nature and culture.

The language's ability to transcend barriers, be they geographical, cultural, or linguistic, positions it as a universal tool for seekers of knowledge and understanding. Its historical evolution, shaped by diverse linguistic influences, lays the groundwork for an ever-expanding mosaic of cognitive progress that knows no boundaries.

English's past serves as a foundation upon which a boundless future of cognitive advancement is built. The language's adaptability, rooted in its amalgamated heritage, allows for continuous evolution and innovation, fostering an environment where collaboration and intellectual exchange thrive uninhibited by borders or limitations.

This interminably ripe future of cognitive progress through English embodies a vision of collaboration that transcends geopolitical boundaries. It fosters a global community of thinkers, scholars, and seekers of knowledge united by a shared language that acts as a conduit for the free flow of ideas and insights.

Moreover, the success of English is a testament to the harmonious synthesis of nature and culture. It draws from diverse linguistic roots, mirroring the intricacies and dynamism of nature, while also reflecting the cultural nuances and richness of human experiences woven into its fabric.

In essence, English's success story lies in its ability to transcend limitations and foster an interminably ripe future of cognitive progress through borderless collaboration. It stands as a spectacular achievement that harmoniously blends the forces of nature and culture, creating a platform where the pursuit of knowledge and understanding knows no bounds.

English transcends limitations to serve all seekers of understanding - its past is prologue to an interminably ripe future of cognitive progress through collaboration without borders. A spectacular success of both nature and culture.

English: The Language of Science and Innovation ©

Conclusion

The English language has been shaped by the influence of Latin, Greek, and Germanic roots, resulting in a language that is both precise and expressive. The blending of these linguistic traditions has created a language that is capable of adapting to a wide range of contexts and disciplines, and has enabled English to serve as a conduit for the exchange and evolution of ideas across diverse realms of human inquiry.

The significance of understanding the historical and cultural contexts in which language evolves cannot be overstated, as language reflects and influences human culture. The evolution of English is a testament to the power of language to shape the modern world and foster global communication and collaboration.

The versatility and power of English have made it a lingua franca, a common language that bridges cultural and geographical divides, facilitating the exchange of ideas and knowledge across borders. The language's ability to adapt to new contexts and disciplines has allowed it to remain relevant and effective in the modern world.

The English language is a unique and dynamic linguistic entity that has evolved over time through the blending of Latin, Greek, and Germanic roots. Its versatility and power have made it a vital tool for communication, commerce, and creative expression, and its influence can be seen across diverse realms of human inquiry.

The roots that have contributed to English - Latin, Greek, Germanic - have each enhanced the language in significant ways. Latin bestowed precision for scientific and scholarly works, while Greek enriched philosophical and mathematical vocabulary. The Germanic foundation lent emotive qualities and narrative abilities.

Where some languages are limited to specific contexts, English has remained highly adaptable through an inclusive approach to linguistic change. By integrating influences from extensive cultural contact, it has continually acquired new terms to describe emerging concepts and technologies.

This flexible nature is a key reason English has ascended as a preeminent global lingua franca. Unlike other languages tied closely to a single nation or region, English acts as a bridge across borders with a broad semantic purview. It facilitates international cooperation in fields as diverse as healthcare, Academia, business and politics.

Looking to the future, English shows no signs of abandoning its syncretic evolution. As digital platforms further globalize exchange of ideas, the language will likely serve an ever-important role in universalizing human learning. Its unparalleled history of cultural synthesis positions English uniquely to cultivate understanding in our interconnected world.

The trajectory of English appears to embrace its syncretic evolution, solidifying its role as a global lingua franca. In an era where digital platforms continue to catalyze the globalization of idea exchange, English is poised to play an increasingly pivotal role in universalizing human learning and fostering global understanding. Its unparalleled history of cultural synthesis positions English uniquely to navigate and cultivate comprehension in our interconnected world.

The digital age has ushered in an unprecedented era of connectivity, facilitating the rapid dissemination of information and ideas across borders. English, with its established presence as the language of the internet and global communication, stands as a natural conduit for this exchange. As digital platforms further globalize interactions, English is likely to become even more intrinsic to the universalization of human learning, transcending geographical constraints and facilitating a shared knowledge pool accessible to individuals worldwide.

The language's syncretic evolution, rooted in its amalgamation of diverse linguistic and cultural influences, equips it with a unique ability to bridge cultural divides. English's historical journey, shaped by contributions from various civilizations, positions it as a language capable of accommodating and integrating diverse perspectives, fostering a more inclusive understanding among its speakers.

Furthermore, English's adaptability in absorbing new vocabulary, idioms, and expressions from different cultures allows it to evolve alongside the ever-changing landscape of human interaction. This adaptive nature positions English as a living language, constantly evolving to reflect the dynamic interaction of global experiences and knowledge.

As the world becomes increasingly interconnected, English stands as a language poised to transcend boundaries and facilitate a shared understanding among people of diverse backgrounds. Its role in universalizing human learning will continue to expand, providing a platform for the exchange of ideas, fostering cultural empathy, and promoting a more interconnected and understanding global community.

English: The Language of Science and Innovation ©

English: The Language of Science and Innovation ©

www.ingramcontent.com/pod-product-compliance
Lightning Source LLC
Chambersburg PA
CBHW082133290526
45794CB00008B/3023